"This guy who I interv
how to please a woman th
else alive. He truly is

~ David DeAngelo, doubleyourd

"I read your book 'Secrets of Female
Sexuality' last evening and I can't thank you
enough for the valuable information. It just
blew me away... finally someone understood me.

"I truly believe this book is the key to a
successful loving partnership. If everyone
out there would read this book, all the
therapists would go out of business; no one
would need them anymore.

"This should be a mandatory read for both
Women and the Men we so dearly love and
appreciate when they are being MEN and
allowing us to be WOMEN!"

~ Jean, Maryland

"I started reading your book 'The Secrets Of
Female Sexuality' and I had to continue until
I finished it. You are right on the money!!!
The 4 things a woman wants is so true. When
someone asks what I am looking for, I tell
them just that.

"The problem is, the majority of men think
they are giving you that and they are sooooooo
far off. I am definitely going to recommend
this book to my male friends as well as my
female friends.

"The reasons you describe why a woman goes
outside her marriage really hit home. They are
all so true.

"I truly believe every guy who seriously
wants to have a great relationship with a
woman read this material. All 3 of your books
are very well written, and should be
recommended to anyone who is interested in a
fulfilling relationship. Heck, all men should
be 'required' to read this stuff. Sex

therapists should be sharing your name with their clients. Had any of my partners had this information years ago, my life sure would have turned out differently!!

"This stuff is awesome! You did a wonderful job at putting this together and I will definitely spread the word!! Thanks!! Keep up the great work!!"

~ Cindy, Michigan

"As a woman I think that in fact what you are doing is a public service for women everywhere."

~ Val, Orange County California

"David, all bookstores in the United States should throw away every book on sexuality for straight men and hang a sign with a link to your website. No man will ever need anything else."

~ Jay, Boston Massachusetts

"Thanks David, for taking this subject out of the garb of mystery and teaching it to normal guys to help them on the road to fulfillment in their lives."

~ Scott M., Ireland

"I say again, you are the voice in the wilderness and the best source of information and encouragement for us guys who enjoy giving women incredible orgasms. Keep up the great work."

~Jim G., Georgia

The Secrets of Female Sexuality

Unapologetic Brutally Honest Truth About Sex
That Women Secretly Wish You Knew But Can't Tell You

David Shade

The Secrets Of Female Sexuality
Unapologetic Brutally Honest Truth About Sex
That Women Secretly Wish You Knew But Can't Tell You

Published by David Shade Corporation

David Shade and Masterful Lover
are registered trademarks of
David Shade Corporation

ISBN 978-0-6151-5807-5

The Agreement

The information in this program is for entertainment purposes only. It is not to be taken as medical, legal, or personal advice. You assume full responsibility for the consequences of your own decisions and actions. Neither David Shade Corporation nor the author of this book will be held liable in any manner whatsoever stemming from your use of the information in this program.

You must always practice safe protected sex. There are other books and resources from qualified medical professionals to teach you how to do that. This program assumes that you always practice safe protected sex as directed by qualified medical professionals.

This program discusses highly controversial sexual activities. Neither David Shade Corporation nor the author of this book assume any responsibility for the exercise or misuse of the topics described herein.

There is a real and absolute distinction between explicit consensual acts between consenting adult partners for their mutual pleasure and all acts of violence against unconsenting partners. Imposing any sexual activity on a reluctant or unwilling partner (or anyone who cannot give legal consent) is a criminal offense. Further, state laws vary: some sexual activities, even between consenting adults, are illegal in certain jurisdictions.

By reading this program you agree to all of the agreement on this page. If you cannot agree to this agreement, do not read this book and immediately return this book for a full refund.

If you do agree to the agreement on this page, then read on, play nice, give women incredible pleasure, and enjoy...

Table Of Contents

SECTION 1.. 1
Introduction... 3
It Was A Very Dark Day.. 10
The Long Road To Enlightenment.............................. 13
How I Ran A Phone Sex Line For Women To Call.............. 17
Does Size Matter?..23
Everything That Society Told Us Is A Lie.............................27
Are Women Sexual Creatures?.................................... 35
What Women Secretly Desire...................................... 41
    Don't Ask Women... 41
    What Do Women Want? It Is An Age Old Question...................41
    What Do Women Actually Respond To?..................................... 43
    There Are Four Things Women Must Have............................... 51
    If Only She Could Have All That She Needs............................. 51
    A Good Man Is Hard To Find..52
The Dark Secret Of Women..58
Do Women Have All The Power?................................62
Passion...64
Romance... 69
Fantasy.. 71
The Rules Of Engagement..73
The Secret To Giving Women Wild Screaming Orgasms..... 82
SECTION 2.. 87
How To Bring Out Her Inner Animal...................................... 89
    The Better It Is, The More She Has To Have............................ 89
    Unleash Her Sexual Potential... 90
    Feel Like A Man...90
    Basis of All Interaction Between A Man And A Woman............91
    The Correct Motivation And Intent...................................... 91
    Genuinely LIKE Women... 91
    Command Respect.. 92
    Earn Her Trust.. 92
    Be A Man...92
    Passion And Absolute Conviction....................................... 93
    Have A Plan..93
    Romance.. 94
    Emotional Connection.. 94
    Calibration.. 95
    The "It Just Happened" Clause...95
    Patience..96
    Dominant..96
    The Remote Control Egg..97
    Show Her How Much She Turns You On..................................... 97

The Importance Of Naughty.. 97
Dirty Talk...98
Phone Sex.. 98
New Orgasms In New Ways..99
The "Yes" ladder.. 99
Fantasies And Role-Play...100
Expand Her Sexual Experiences...100
The Power Of Possessive...101
Piece Of Property...101
Total Surrender.. 101
Post Coital Re-Bonding... 102
To Be A Masterful Lover... 103
SECTION 3.. 105
Success Stories.. 107
SECTION 4.. 135
Resources...137
Get Your FREE CD Today!... 140
FREE VIP Inner Circle Offer... 141
FREE Stuff Linked To This Book..142

Preface

For the ladies; while this book is written in the narrative of me writing to men (for obvious reasons) I want to include you as well.

For the men; **WARNING!** I am going to present radical changes in beliefs. Much of what you are about to read will be contrary and challenging to everything you have believed about female sexuality, and about relationships altogether. Some men have condemned it even before reading past the first chapter. (You will read e-mails from two such men). I have coached thousands of men, and invariably it is the ones who reject the truths of female sexuality who are the ones who regularly find themselves alone, or worse, stuck in a very unfulfilling and unhappy marriage.

You may find that the information makes you feel uncomfortable. I went through a very painful experience as a result of misconceptions and ignorance of truths which proved to be the beginning of a journey of discovery and personal empowerment (which you will read about). Positive change is invariably accompanied by discomfort at first, but I intent to save you from the kind of pain that I went through. You have to keep an open mind.

In this book, I unapologetically present the truths of female sexuality in a brutally honest manner. Most of the other books you find in the bookstore are written by 'feel good' politically correct theorists who sell to you by telling you what you want to hear and things which are in agreement with social programming. That stuff is not based on real cases from the real world, and is of no help in situations where your woman is unhappy and not having orgasms (usually going hand in hand).

I tried all the books in the bookstores. Very few had anything truly helpful. The ones that stand out are books such as "The Hite Report" which is 440 large pages of small print of factual reality attesting to the fact that women are unfulfilled to the point of utter madness.

I went outside the boundaries of political correctness and societal norms and tried things not done before and found the things that really work. I did the actual field work and found what women actually responded to. Nothing else matters except how women respond.

Most of what you have learned about female sexuality was probably from people who have misconceptions themselves. Looking at it more closely, you will probably find that they have frustrations of their own. Unlike them, I have had success that is known worldwide. My clients have repeated these successes to give their women indescribable pleasure beyond what can even be comprehended by most people. (You will read from some of them). Hence it is no surprise that my thesis is in contradiction to what other people may have told you before.

On another point, I suspect you may be quite skilled and successful yourself. Most of the men who read my materials are into learning to be better lovers themselves, and thus are already far ahead of most men. Many of my top clients were very impressive themselves, yet became even more so. If you are such a man, I would hope to hear from you, as we would have much to talk about.

How To Learn More On A Continual Basis – Free

I strive to see that the truths in this book work for you, and I hope for you to use them. In that light, I have a number of real world cases from other readers, answers to questions, advice to challenges, and very useful tips in my free e-mail newsletter. To get your copy, you need to go online to RenegadeSexExpert.com to sign up.

Book Roadmap

Section 1 begins with a key turning point in my life, the sequence of events that began to crumble my misconceptions of sexuality, and how I began to discover the secrets. As misconceptions were falling apart, they began to be replaced by my field work findings of how women actually respond. Nothing matters except how women respond. Then by trying things that were way outside the boundaries, you will read how I made things happen that are beyond the comprehension of most people. As I outline what is really true; the secrets are fully developed and explained. I also include the key points that I learned from other authors that I respect. I then share how to apply the secrets in the noble cause of giving women wild screaming orgasms.

Section 2 is a summary of the things you must do in applying the secrets in the supreme endeavor of awaking a woman's sexuality. These are the key points. They are covered in brief such that they may fit into this book and such that you may begin leveraging the secrets right away. For the interested reader, I refer you to my other works which cover the entire topic in detail including many specific detailed examples.

Section 3 is the real life true success stories of just a few of my clients. These serve as good examples of applying the secrets to make women delirious with pleasure, wildly happy, and totally loving everything about being a woman.

Section 4 is a list of resources for further discovery and development. I also include the list of books and mentors that have affected me along the way. I have consumed countless books and resources on this topic, and can recommend the few that contributed leading work. Then free offers and resources. These include a free CD, two months free trial membership to the Insider's Circle, and free website resources which includes my free e-mail course.

SECTION 1

The Secrets

A woman is a mystery to be solved. But a woman holds nothing from a true lover. Like the blush of a rose, pink and pale, she must be coaxed to open her petals with a warmth like the sun. It calls for the lust of a whale crashing to the shore, so we may steal up what lies beneath, and bring the foamy delight of love to the surface.

- Don Juan DeMarco

Come then, Come then! Give me thy hand, oh fairest! Whisper a gentle 'yes.' With joy my life thou'lt bless. Come, dearest, let me guide thee. Danger shall ne'er come nigh thee.

- Don Giovanni

Introduction

First of all, let me explain that all of this information only pertains to mentally healthy women. In fact, it best fits women who are intelligent and have high self-esteem.

And when I say "women this" and "women that", I mean MOST women. Certainly there are exceptions to everything.

And this only applies to women who you intend to see on a continuing basis. It does not apply to one-night stands. The most important reason for this is that only women who take you seriously are going to respond to this.

Remember, when you lead a woman, you are responsible for her. As they say: "With great power comes great responsibility."

If you are familiar with my material, you know that it can at times get hard core. If you are not familiar with my material, some of it may seem a bit shocking. But let me make this perfectly clear: all of this is in the context of respect for women. I have the utmost respect for women. My parents raised me right.

Also, this is NOT about manipulation of women. I loathe when men try to manipulate women. Instead, this is about empowering your woman. It is about "facilitating", about "enabling".

But at the same time, I am all about results, and sometimes that takes drastic action. Let's take a very tame example.

Let's consider the case in which you are with a woman who has difficulty having orgasms...

I recommend that you NOT tell her that the goal is for her to have an orgasm. It may be the honest truth that you want to give her an orgasm, but to tell her that is only going to make things worse, because it gives her "performance anxiety".

Therefore, it could be argued that not telling her that you intend to give her an orgasm is not telling the truth.

Well, in this case, to tell her the truth is NOT helping her.

Instead, you simply tell her that you want to learn what feels good for her, you tell her that you simply want her to feel the pleasure, and you tell her to do exactly what you instruct her to do. Eventually, as a result, she has an orgasm!

Is that being dishonest? Is that withholding the truth? Is that manipulation?

I think that it is "facilitating" her. It has empowered her.

That was a very tame example. There will be other more powerful situations, but just because my methods are powerful, and actually work, does not mean they are manipulation. My methods are "enabling".

Also, absolutely, this is NOT about having power over women! Some people get that wrong.

Let's consider an e-mail from one person:

    I was expecting something different from
    this book. Please issue me a full refund.
    After reading the first portion of this book I
    feel like you have chosen to seek sexual power
    over women to cover up the fracture of your
    past. You are not yet healed. I hope you are
    open and vulnerable enough to continue seeking
    help, no matter how good you are in the
    bedroom. Area B does not fix the pain of area
    A.

That customer immediately received a courteous and complete refund. Still, it's too bad he didn't read the rest of the book. He would have read about the very points that he mentioned...

"was expecting something different"

I don't tell men what they want to hear. If he had read on, he would have read what I tell men that they need to hear.

"seek sexual power over women"

If he had read on, he would have learned why that is the wrong thing to do in every way. He would have learned the concept of being Personally and Sensually Powerful, and the concept of empowering women.

"cover up the fracture of the past"

He would have read how to leverage the past to bring you to an empowered future. He would have read about how I assumed responsibility for my lot in life, and how I became empowered to do something about it.

"not yet healed"

He would have read the full story of the healing process, including the mentally healthy way to healing based on the critical ingredient of personal responsibility, how that leads to self-empowerment, and why that builds self-esteem.

"open and vulnerable"

He would have read about the importance of being open and vulnerable, and its magic in the context of being strong and secure. He would have understood how that integration is a truly masculine virtue, and exactly why women are so drawn to it.

"continue seeking help"

He would have read about the importance of continuous self-improvement, and how others can help. In fact, he would have read how women helped me so very much in my journey, in ways they didn't know, but in such important and key ways.

"no matter how good you are in the bedroom"

He would have read how being good in the bedroom is not a compensation, is not a cover up, is not a magic pill. He missed out entirely on the resultant understanding that being good in the bedroom is merely a byproduct of all the other things.

"Area B does not fix the pain of Area A"

That is entirely congruent with my book.

I also want to say that I think that men and women are equal. I do not believe that men are superior to women. I do not believe that women are subservient to men. My parents raised me better that that.

Here is an e-mail from a man in Sweden:

I like your work and have bought all your products. I did some of it before I came across your work and I felt really good when I read your books. I understand that they are not politically correct, especially in my country (Sweden) where women and men are considered equal in every aspect, but I know you are right.

I believe that as well. My mother has always been a woman who commanded respect and insisted on being treated equally. (Note: she was a Daddy's girl. That is a good thing.)

She and my father have always had a wonderful relationship.

They raised my sister to also command respect, which she most certainly does. (Note: she is also a Daddy's girl.)

My parents raised me to treat people with respect. They taught me that I am accountable for my own actions. I have them to thank for a loving, functional, and healthy upbringing.

But to be totally pragmatic and realistic, for the sake of a successful and exciting sexual relationship between a man and a woman, there are certain everlasting truths that must be recognized.

Women are feminine. Men are masculine.

Women like being women and like the contrast.

Men like being men and like the contrast.

Part of being feminine, for most women, is to be sexually submissive.

Part of being masculine, for most men, is to lead sexually.

For the most part, women experience their sexuality in the context of the man leading.

More about this later.

Still, the question remains: is that equal?

It is not exactly "equal", it is instead "complimentary". The important thing is that it is consensual assumption of roles in the context of mutual respect.

Interestingly, that's the situation in which most couples experience fulfilling sexual relationships.

But some people get all the wrong idea about "The Secrets of Female Sexuality". Some men see women as the obstacle to get around to get to the sex. That is all wrong.

Also, this book is not about promiscuity. Let's consider an e-mail from another person:

```
    What  a  fool  I  am  to  get  scammed  again  by
some  sleezy  two-bit  creep  like  you.   For  some
barhopping,  phone  sex  operating  predator  to
think  he  has  any  idea  what  a  "quality"  woman
is  about  is  insane.   Okay,  so  you  are  really
good  at  getting  in  the  pants  of  slutty,
cheating  women  who  are  out  looking  for  someone
to  get  into  their  pants.    Has  it  never
occurred  to  you  that  most  people  are  not
promiscuous?   The  people  who  aren't  are  not
spending  any  time  at  your  bars  or  on  your
phone  sex  line.   Your  bullshit  is  not  a  whole
lot  of  help  to  a  family  guy  who  wants  to  be
better  at  loving  his  sweet,  tender,  non-
cheating  wife.
```

That person's purchase was promptly refunded.

This book has absolutely nothing to do with promiscuity. This book is about the truths of female sexuality.

By the way, women are not out looking for someone to get into their pants, at least the high quality women. Again, all of this is only in the context of high quality women.

When I use the term "slut", I am referring to it in a good way.

It does NOT refer to women who are promiscuous, or seek validation from sex, or sell themselves out. Most men are not interested in such women.

It refers to women who want to share their sexuality with their man, and celebrate their sexuality and enjoy it in a ruthlessly expressive, animalistic, natural way. That's what we all really want anyway.

And I most certainly aim to help a family guy who wants to be better at loving his sweet, tender, non-cheating wife. I am all about that.

Here is an e-mail from such a man:

David...

I read your e-mails with interest, and was concerned enough to pen my own when I read the the one from the "family guy with the non-cheating wife".

I've bought both "David Shade's Manual" and "Give Women Wild Screaming Orgasm" books, and my wife of 10 years has benefited greatly. I often surprise her with "new stuff" now, and have used the Welcomed Method and Deep Spot to take our sex life to a whole new level. She loves it!

I've benefited by no longer being so selfish in this aspect of our marriage. I now serve her, and she in turn desires to serve me more, so it's more mutual now than in the previous nine years of marriage.

And our children benefit because my wife and I are communicating better than ever and are more passionately in love.

What you've done for my marriage, David, has been amazing! You've taught me things I've never heard of before and helped make my married life far more compelling, enjoyable and fulfilling.

Dave (another family man)

With that aside, I begin with my story. I was married once ...

It Was A Very Dark Day

Though it was a beautiful sunny Tuesday evening in April of 1992, it was to be a very dark day for me...

Returning home from work, I pulled into our subdivision of large brick homes. Golden beams of sunlight shone through the trees onto the carefully manicured lawns.

I pulled into our garage and went into the house. I hugged my two children and I kissed my wife.

She was hot. A lingerie model. 5'9", 36-24-36, beautiful breasts, amazing ass, long athletic legs, long dark hair, and full red lips.

I had met her 10 years before. We had sex on the third date. Two years later, we were married.

We had built a big home in the suburbs. We had two beautiful children. She and I had sex every night. It was basic vanilla sex, but we always had simultaneous orgasms. I was a happy man.

Recently, though, the frequency of sex had diminished. We had started arguing about money. It got to the point where I felt that I couldn't do anything right. I didn't know what was wrong.

I asked her to sit on the couch with me. I pleaded with her to tell me what was wrong. She simply replied, "I want a divorce and you get the kids. I'm moving out." I was floored. I asked her to explain why, but she refused.

Later that evening I was in my den. Behind me was a wall full of Mechanical Engineering textbooks from my undergrad and Electrical Engineering books from my Masters. Before me was a large solid oak desk that my parents had given me for a housewarming present. They were to be the only things that I would walk away with after the divorce.

I picked up the phone to make a call. I heard my wife on the extension in the bedroom telling her sister about the man she had been fucking for the last two months.

I quietly set the phone back down and I went out my front door to sit on the steps. I stared into the lawn. It felt like I had been punched in the nose and my heart ripped out. It hurt like Hell.

I had treated her like a queen. I was a good provider. We had good sex. Why?

The house went up for sale. The kids and I moved into an apartment. Every other weekend the kids went to her apartment.

Sometimes on the weekends that I had the kids, I'd get a baby sitter and I'd go out to the nightclubs with a buddy from work.

One night I saw her there. She was with the guy she'd been fucking. They were all over each other. It was disgusting.

He was a classic Bad Boy. More precisely, a machismo asshole. He had all the smooth moves.

On a weekend that my wife had the kids, I went to the nightclubs, and I saw that guy there. He was with another woman. He was doing to that woman what he had done to my wife.

He was startled when he turned to find me standing right in front of him. I assured him that I was not there to kill him, but simply to ask him something. I asked him what he had done right and I had done wrong.

Perhaps out of fear for his life, or perhaps to brag, he started telling me everything.

He spoke in terms of examples. For educational purposes, I have distilled the morals of the examples here:

1) Find out what she needs and give it to her.

2) Women need constant reassurance, but at the same time, constant doubt.

3) Never let her think she "has" you.

4) She must always be jealous of you.

5) Always remain interesting and challenging to her.

6) Always keep strife in the relationship.

7)Be possessive of her. She must feel "owned".

8)Never let her look at another man.

9)Keep other men away from her.

10)Show her more excitement than any other man has.

11)Sexually fulfill her, and then some.

12)Never let her feel she fully satisfies you.

13)Know her weaknesses and play them when needed.

Then he said, "I'm glad you're not the jealous type." I responded with, "If I were, you would not be here." He ran, because where he comes from, such matters are dealt with in very violent ways.

The following week I had to drop off some papers at my wife's apartment. As I was about to knock on her door, I heard her and him having sex inside. She sounded like she was enjoying it a lot more than she had enjoyed sex with me. I listened until I heard her come very vocally, and then I slid the papers under the door and left.

Never again would a Bad Boy steal a woman away from me, and never again would a woman be anything less than completely fulfilled and breathless after sex with me.

The Long Road To Enlightenment

I got my wits back and settled down. No longer was I interested in revenge. I simply wanted to be better and live a fulfilling life.

I knew what that Bad Boy was doing when he worked the women. I wanted to have the same effect on women, but I didn't want to do it by taking advantage of a woman's vulnerabilities. I wanted to do it by empowering her strengths.

I started by reading divorce recovery books. I read as many as I could.

That got me interested in books about relationships. I read tons of them.

It started to become apparent to me that my relationship with my wife had been doomed from the beginning. Basically, as it turns out, she had low self-esteem.

Even though she had the perfect life, and was treated like a queen, she did not have the sense of deservedness enough to appreciate it. In fact, it was against what she believed that she deserved. She honestly believed that she deserved to be treated poorly. And thus, she sabotaged what she had with me.

So, basically, it was all my fault! I had chosen poorly!

That's right. I had only married her because she was so smoking hot.

That was the first step to recovery. I assumed responsibility for my lot in life.

Because of what I had been through, I worked hard to devise a way to choose the correct women, which has since served me well.

At the same time, I realized what that Bad Boy was doing. He was preying on the insecurities of low self-esteem women.

"Women need constant doubt." "Never let her think she has

you." That will certainly keep a low self-esteem woman on her toes. Low self-esteem women are always chasing what they can't have. But a high self-esteem woman will tire of that. A high self-esteem woman wants to get to the point where she can enjoy having her man. If she can't, she'll move on.

"She must always be jealous of you." True for preying on the insecurities of low self-esteem women. Very bad for high self-esteem women. There are much better ways to have the occasional "healthy" jealousy that keeps a high quality woman sharp.

"Always remain interesting and challenging to her." Agreed. However, there are healthy ways to do this.

"Always keep strife in the relationship." Wrong! That will attract low self-esteem "drama queens". It will repel high quality women.

"Be possessive of her. She must feel owned." Very bad! High quality women do NOT want to be "owned". I will later explain the correct "occasional" context in which this is actually a good thing, and very powerful.

"Never let her look at another man." "Keep other men away from her." Bad Boys are very insecure paranoid men who know that other Bad Boys are always trying to steal their woman away. In actuality, such a mind-set will repel high quality women. In reality, there is no need!

"Show her more excitement than any other man has." "Sexually fulfill her, and then some." Agreed. Most definitely. I have even gone MUCH further!

"Never let her feel she fully satisfies you." WRONG! In fact, for high quality women, I will show you that in fact you want to show her just how MUCH she satisfies you, and why that is so powerful!

"Know her weaknesses and play them when needed." That is just so fucking stupid! Typical of the low self-esteem Bad Boy.

I have devised much more rewarding ways to keep a high self-esteem woman excited about the relationship, and in fact, to bring out her inner SLUT!

Since I was back in the dating scene again, I wanted to be a memorable lover for the women that I was meeting and I wanted to enjoy the experiences myself to the maximum extent possible. So I read all the books about sexuality and being a better lover.

Unfortunately, they were somewhat disappointing.

On the weekends that the kids were at their mother's, I would go to the nightclubs.

At the time I was 36, and felt old and used up. Then, I began to discover that some of the young women found me attractive. I started dating a 19 year old woman who had gotten in the nightclub using a fake ID.

Unfortunately, she was not having orgasms. I was making love to her like I had made love to my wife for many years. I realized soon that that was not working on the new women.

Later she started dating a Bad Boy. One night she and I ran into each other and I asked her how things were going. She was happy to report that her new boyfriend had given her her first orgasm!

Damnit! A Bad Boy had beaten me again!

I decided that that would never happen again.

I read more books on sexuality and being a better lover. I purchased educational videos. I listened to audio recordings. There may have been one paragraph in each book worth quoting, but otherwise they were all the same old stuff rehashed.

I decided to try stuff myself and figure out what really works.

I started dating a 22 year old woman. She also had never had an orgasm. She had only been with one man before. They had been together for a year, and no doubt he had tried everything on her.

So I just took my time finding what would feel good for her. With my middle finger I searched for her "G-spot", like all the sexuality books recommended, but I received no response from her.

Then I decided to slide my finger in as far along the front wall of her vagina as I could. I curled the tip of my finger and pressed hard against the front wall. She immediately responded to this.

After continuing this for a few minutes, I watched the face of this beautiful woman while she experienced her very first orgasm.

It was beautiful to watch. It was wonderful to be there, to be a part of it, and to have helped to "facilitate".

I decided that I was going to be my own sexuality teacher.

I gave more women their first orgasm.

I converted women who had never masturbated in their life into masturbating maniacs.

I turned women who insisted that they could only have one orgasm into multi-orgasmic come machines.

Interestingly, I later came to realize that I had actually benefited most from the education I received when I was writing my masters thesis in Electrical Engineering on Artificial Neural Networks.

As computer engineers, we looked to the human brain as an alternative computing architecture. It did things seemingly effortlessly that we could not get computers to do.

We read papers by neurologists. They quoted papers by psychiatrists and psychologists. That got me very interested in psychology.

I also learned some neural physiology and was starting to form an understanding of how the brain processes sexuality.

So I started to look at the power in the psychology and neural physiology of female sexuality. That's when things started getting very interesting...

How I Ran A Phone Sex Line For Women To Call

After my divorce in 1992, I felt very defeated and alone. I had custody of our two small children, and thus in the evenings I was stuck at home. Hence, after I put the children to bed, I turned to the telephone.

There was a singles magazine in our area where people put in personal ads. You would read through the ads and decide which women to call. You would then call a 900 number and leave a message and your number. The ladies would then listen to their messages and decide who to call back.

That was back in the days before there was caller ID. The women could call and know that you had no way of knowing who they really were or where they lived, so there was that anonymity which gave them safety.

I picked women who were recently divorced and about thirty years old, so when they would call, the rapport would build quickly as we had much in common and much to talk about.

I would build common ground, based on our similar situations of having gone through a divorce. This allowed her to feel comfortable with me.

I also became very good at establishing an emotional connection with a woman on the telephone. This is critically important. Women are emotional creatures and they need to establish a connection before they can feel free to continue further.

Since she understood my situation, I would say to her, "I really enjoy talking to you. I feel like I can tell you anything, and you understand it and accept it." She would reply, "Oh yes, I do!"

This served to make her feel that she could tell me anything as well and that I would fully understand it and accept it. Thus, they would open up even further.

For many of these women, by the end of their marriage, they no longer felt sexual. They weren't having orgasms, even by masturbating. They didn't even fantasize. There was no point in it.

They had left their marriages for various reasons, but in all cases the sex had become boring. Their husbands were lousy lovers. I asked these women, "Was it because he had a small penis?" They would reply, "Well, actually, no."

Many of them had affairs with exciting lovers. They talked about how much their lover turned them on. I asked these women, "Was it because he had a big penis?" They would reply, "Well, actually, no."

Apparently, it didn't have anything to do with the size of the man's penis.

All of my preexisting beliefs about sexuality began to crumble.

All of this so far may sound as if I was being very pathetic, and as I look back on it, I can see it that way.

However, hearing these womens' stories was so fascinating to me.

It was also therapeutic for me. It was sort of a "mutually helpful divorce recover program". It helped me to deal with my recent divorce. Also, I have to admit, it helped with the loneliness.

The women were lonely too. They were reaching out in the dark to talk to another human being, to have thoughtful human interaction.

For them it was also therapeutic to share their secrets, even with an anonymous person, and because it was anonymous, they were completely open and honest.

I was learning a lot. The stories were very revealing, and quite fascinating.

I became very good at getting them to open up and share.

I would ask them, "What do you really want in a man?" They would describe that, or at least try to describe it to the best of their ability. Often they didn't really know themselves; they just knew what they wanted to feel. The important thing is that they started to feel those feelings while they were on the phone with me.

So I would then ask them, "How would that make you feel to be with such a man?" They would begin to describe that. In order to describe it, they had to imagine feeling it. This caused them to actually begin to feel those feelings and imagine that they were really with that man, and because they were talking to me, and had rapport with me, those feelings got associated to me on a subconscious level. They did not realize this on a conscious level.

Because I was genuinely interested in what they thought, and because I made it a point to demonstrate that I was a good listener, they opened up about what they would really like to have. In other words, what they fantasized about on an emotional and relationship basis.

When they were basking in a flood of good emotions and feelings, which were all connected to me, I would ramp that up. After having established an emotional connection with them that allowed them to feel very close to me, I leveraged that to move the conversation in a romantic direction.

I would have them describe to me what they would do on the ideal "date". They would describe some romantic night out consisting of dinner and dancing and walking on the beach. Of course, all of this would be described in detail. "Then we are seated at our table, which is adorned with a fine white linen tablecloth and one tall candle."

Sometimes they would instead prefer that I describe the ideal date. If so, I would simply narrate the ideal date as described to me by previous women. Simple really. Give women what they have to have.

After they are basking in the flood of romantic feelings, all of which is linked to me, allowing them to feel romantic towards me, I would then covertly move things in a sexual direction.

I would say, "Do you ever feel alone?" They would affirm.

Then I would pose the question to them, "What would it feel like to be the opposite of alone? What word could be used to describe that?"

They would try to describe it and try to put a word to it. It served to remind them that they are alone and to remind them how much they yearn to feel close and connected.

Then I would say, "Yeah. Sometimes I feel alone. Especially at this time of night. I think about what it would be like to be with that someone special. What would we say? What would we do? How would we make each other feel?"

They would softly say, "Yeah..."

Then I would say, "I feel very close to you right now." They would reply, "Yes, I feel very close to you too."

Then I would softly say, "I wish I were there with you right now." They would softly reply, "Yes..."

"If I were there right now, I would want to hold you so close." They would softly reply, "Yeah..."

"And I would feel your soft skin against mine." They would sigh.

Here is the critical point...

"And I would ever so softly kiss the side of your neck."

At this point, fully 80 percent of the women would simply stop talking and just sigh.

I would continue...

"And I would softly kiss the side of your neck all the way down to your shoulder."

I would slowly describe in detail everything I would do if I were there, beginning with what I would kiss. Then what I would touch.

They would start moaning.

Then what I would lick.

Within minutes, these women would be screaming in orgasmic ecstasy.

Fully 80% of these women, all attractive and educated, on the very first call, to a man they have never talked to before in their life, would engage in phone sex with me.

Have you ever heard of that before? Intelligent, educated, professional, successful, attractive women calling a strange man and having phone sex?

It's usually the other way around, where men pay $4.99 a minute to call a woman to have phone sex. Besides, that woman is actually an ugly old hag in a hair net smoking a cigarette with a monotonic foreign accent.

I continually adjusted and refined my phone sex techniques, making it progressively more realistic and effective. Just from my voice and the reality that I created in their mind with sensory rich descriptions and erotic words and eventually outright naughty vulgarity, women were having orgasms, then women were having multiple orgasms, then women were having orgasms without even touching themselves.

I became extremely good at giving phone sex. I was absolutely lethal. The phone would ring and I knew that within one hour another woman that I had never talked to before would be screaming in orgasmic ecstasy.

It became routine. I had to work the next day, and I needed a freshly ironed shirt for work, so I'd be getting women off on the phone while I was ironing my shirts. I would literally be having phone sex while I did my laundry!

Picture this... I am looking down at my ironing board, holding the collar down of my white shirt with my left hand while with my right hand I press the iron across with the phone pinched between my right shoulder and chin while I say, "Keep coming for me Baby! I'm gonna come! Now! NOW Baby! Come harder!" Meanwhile the woman is screaming at the top of her lungs right into my ear.

I was leading a private life in the evenings running a phone sex line for women. I started feeling like a sex industry worker. I should have been wearing a hair net and had a cigarette dangling from my mouth with a really long ash on it that's about to fall off onto my shirt while I squint my eyes to avert the rising smoke.

These women were well-educated professionals. Elementary school teachers, mid-level managers, emergency room nurses, stockbrokers, sales directors, fast-rising corporate executives, you name it. They all had high self-esteem. They came from good homes. They were the farthest thing from sluts. When I got on the phone with them, however, and unlocked their pent-up sexual potential, they became total sluts.

These women really opened up and taught me things that fascinated me. They spoke of their sexual past and their fantasies. I learned a great deal about the secret sexual lives and thoughts of women from all of those conversations. That's when I really started to understand the underlying sexual potential that exists within women.

My beliefs about women began to be shattered and totally rewritten. I became even more fascinated with the minds of women.

Women are not that complicated after all. It's just that we don't understand them. And that is completely understandable seeing that social conditioning has lied to us. Even some of the so-called "sex experts" had unintentionally misguided us through their own ignorance or political correctness.

I am trying to get you to begin to understand women and to begin to rewrite your beliefs. When you truly begin to understand women, they are no longer intimidating, but instead, they are absolutely fascinating and thoroughly enjoyable.

The interesting thing is that even at that time, during my "phone sex period", I began to understand that what I was doing was something very novel. I was onto something that nobody knew about. There was nobody that I could talk to about this, because nobody could even comprehend that it was even possible.

And I had absolutely no idea at the time that I would end up teaching this stuff. I just wanted to learn.

For me, at the time, I was on the beginning of a long learning curve that would prove to be the beginning of a fascinating journey.

The reason that I was beginning to be very good at it is that I genuinely had a passion for it. That is key.

Does Size Matter?

There is the classic belief that a large penis is the key to satisfying a woman.

When those recently divorced women from my phone sex era told me how their husbands were such boring lovers, I asked them, "Did he have a small penis?" They replied, "Well, actually, no." When they told me of their exciting lover that they had an affair with, I asked them, "Did he have a big penis?" They replied, "Well, actually, no."

That is when I first started to question the contribution of penis size.

When the internet started to become popular, I met women online. I met a number of married women who were looking to have an affair because they were tired of getting skewered by their husband's big cock. They found their times with me to be "refreshing", as well as multi-orgasmic.

Even years later, I have ex-girlfriends coming back for more while telling me about their current lovers with large dicks, saying, "But he doesn't make me feel what I used to feel."

Here is an e-mail that I received from a female reader of my newsletter, which is typical of many e-mails that I have received:

```
I had sex with a man that had a medium size
penis and I used to cum like crazy. Now I have
met this other guy that is well endowed with a
very nice size penis, but it is really hard
for me to cum. Can you help me figure this
out?
```

Here are some actual e-mails that I have received from men (who had not yet purchased any of my Masterful Lover products):

I have been either blessed or burdened with a big penis. The woman I am with now has lots of trouble taking in the whole thing.

She said it feels like it's bottoming out most of the time, to a very tender area inside, leaving her with a very uncomfortable feeling.

If in the missionary position, if I slide in slowly, she's ok.

I can't go hog wild because it hurts her, but if I don't go hog wild then I find it hard to maintain an erection.

She has never had a vaginal orgasm. She has only had orgasms through clitoral stimulation.

I've been working with her trying to get her to orgasm in the "G-spot" area, with no success yet.

Do you have any suggestions or positions that would make sex more enjoyable for her and me?

In reply to your e-mail, I would say to you: She is not having orgasms in intercourse, so what's in it for her? The same old routine every time and she never comes. It gets boring, and the regular impaling against her tender cervix becomes downright irritating. It gets really old.

To give her her first vaginal orgasm, use your 3 1/2" long middle finger on her deep spot, which isn't that "deep" actually. It's just "deeper" than the G spot. Once you do that, she'll be much more receptive to intercourse, though you're still going to have to choke up on your bat.

Here's another e-mail (complete with the misspellings):

```
    my girl has never had an orgasim.  i have
had sex with her four miserable times doing
what  i  thought  was  "good"  but  failing
miserably.   How do i change the current from
what was done to what should be done, causing
her to have multiple orgasims and having her
trust me despite my last performance?  i have
a big cock but, this is really humbeling to
write you about my shortcommings, but not only
do i need to change precendence verbally but
more  imporantly,  physically....  going to buy
your stuff bro, need help!!
```

You can tell a lot about a person based on whether they use an upper-case 'I' or a lower-case 'i' to refer to themselves.

Here's another e-mail:

```
    I recently got with a 23 year old female who
told me that she has never gotten an orgasm
from any guy whose penis size is under 10
inches.  I told her that I was an 8 and one
night when we were having sex she said that
she wasn't feeling any orgasms any time soon,
so we stopped.  I felt weird and tried finding
new ways to give her an orgasm.  What can I do
to make her have a mind-blowing orgasm that
will make her forget all about a man's size?
```

In reply to your e-mail, I would say to you: You may have always thought that you were big, but she's got you thinking that you're too small. She's good, and you're being a wimp.

She said that she has never had an orgasm from any guy under 10 inches. You took her challenge even though you didn't measure up. That puts you in the wimp position. You should have simply said, "Sorry Honey, can't help you. I'm only an 8."

Possibly, is the order of things said in your story a little different? I can imagine a man trying to impress a woman by saying, "I have an

8 inch cock," and then the woman cutting him down to size by replying with, "I've never gotten an orgasm from any guy who's under 10 inches." She doesn't take you seriously anymore, and thus she is not going to respond to you in bed.

Frankly, I'm surprised she went to bed with you. She was probably curious to see what it was like to be with an 8 inch cock. Evidently disappointing.

Here's an e-mail from one of my clients:

David, I am having the most incredible sex, as often as I want, and of course, I am giving her the best sex of her life and she can't get enough. Do you want to know the one thing she always asks me? She'll look at me after I've just taken her, or given her an orgasm over the phone, or told her that I am going to fuck her in her office, or anytime I am direct and tell her what I'm going to do? She looks at me expectantly and asks, "You're never going to change, right? This isn't some sort of trick? You'll always be this way, right?"

She had two orgasms the other night before I even got in her all the way. I just stuck the head in and let it sit there and she had two - just from the head.

This is, of course, a hot girl that is used to having men pursue her and wine and dine her, take her on private airplanes to movie premiers, etc, etc, etc.

David, life is good. Life is oh, so fucking incredibly good.

Thanks for showing us the way.

Everything That Society Told Us Is A Lie

Movies, television shows, stories, what other people told us, what our parents told us, what women told us, all that "social conditioning" told us that we had to talk women into having sex. It almost made it seem as if sex for a woman was a chore she performed, an obligation she fulfilled, only as a reward for us spending money and attention on her.

That single misguided piece of social conditioning had me perplexed for years.

For me it was a belief that was hindering me, a self-defeating belief. I was determined to crush it by finding the truth, and then to replace that belief with empowering beliefs based on reality.

I began to realize that all the social conditioning was not jiving with what was actually happening in reality.

During my 20s, most of my girlfriends absolutely loved sex. It's not that they initiated it, but they would hint at it and then the moment I would initiate it, they would jump on it. Then in my marriage, my wife had to have it every single night. From my point of view, it appeared that women loved sex.

So in my phone sex phase after my divorce, I tried a cruel experiment. Some of those women that I talked with on the phone, I met up with in person. We'd meet at a neutral place, have a drink, and then go to her place.

What I would do is I would get her naked on the bed, and then I'd hesitate. I'd stall for as long as I could. If she persisted further, I'd tell her, "I'm not really ready for this yet," or, "I think maybe this is moving too quickly." Basically, I was purposely NOT going to have sex with them, just to see what would happen.

What happened is they got PISSED! Either they'd be insulted and start sulking, or they'd cuss me out. One woman even called me a coward! I made peace with them, got back on their good side, and then obliged them. Then they went at it to such a degree as to make themselves look selfish.

I laugh now as I remember back on it, but it proved something very important to me:

Women Love Sex!

Women actually WANT to have sex! Not as a reward to 'a nice guy', but simply because WOMEN LOVE SEX!

(That does NOT mean that women are promiscuous.)

That, along with the stories that women shared with me over the phone, along with all the factual evidence from reality, clearly showed that not only do women love sex, but in fact women are far more 'sexual' than men.

Further evidence of this can be found in any woman's fantasy book, such as "My Secret Garden" by Nancy Friday. That will open your eyes. Women have elaborate intricate fantasies beyond anything men can imagine. For women, it is largely mental and emotional, and about being "taken" and "ravaged".

So, more accurately...

Women Love To Get FUCKED!

That does NOT mean that women want to be raped. Certainly NOT! Rape can potentially destroy a woman's sexuality for life.

On the other hand, in the context of consent, mental stimulation, emotional rapport, and trust, women love to be taken and ravaged and fucked hard.

I always thought that women only liked to "make sweet love". They certainly sometimes do, as do I, but they sometimes also like to get fucked! So I started doing that, and they got crazy!

I pushed the envelope even further. While fucking them hard, I would start to talk dirty. They got crazier!

During my last two years of college, I had a college sweetheart who I found to be very cute and sexy. She also happened to be very intelligent, very high self-esteem, and extremely sexual.

We had very good sex. We did it often and always had simultaneous orgasms.

One day she said to me, "As you know, I love horses. Well, I also admire how they are very animalistic when they have sex. And when we do it doggy style, I sometimes imagine that I am a horse and that I am being very animalistic."

I couldn't believe what I was hearing, but of course I said, "Oh good Baby, I love that you are able to feel animalistic."

I had certainly heard that women like to be "animalistic". I never really thought that they went as far as to imagine they were an animal!

So, the next time we did it doggy style, I said to her, "Oh yeah Baby, you are being very animalistic!"

She got even more excited. I couldn't believe I was saying this.

I took a big gulp of courage and I said, "You are so animalistic, just like a horse. Whinny like a horse!"

She started making horse noises! She started going crazy!

What do horses do? Whinny? I don't know.

I continued, "Oh yeah Baby, make like a horse! You're getting fucked like an animal!"

She went ballistic. She started screaming in orgasmic ecstasy.

I couldn't believe it. It was so "foreign" to me, but hell, it worked, and very powerfully. Truth is found in reality.

My next girlfriend had no interest in horses, but she did like to be animalistic. I altered what I said to fit the situation. It worked amazingly well.

I was onto something. I varied it further with each subsequent girlfriend. I moved it from being animalistic to being slutty. That worked even better.

I found that the more vulgar and raunchy I got, the more excited they would get.

Women Love Dirty Talk!

I pushed the envelope even further. I would see how very naughty I could make them be.

I would make a fantasy scenario where she is a very naughty little schoolgirl who was caught masturbating, and I am the mean disciplinary principal. I would spank her and tell her how naughty she is and how she needs to be disciplined. "You're being very naughty! You will learn to be a good girl! You are being so naughty, but you want to be a good girl! Come like a good girl!" This would consistently result in powerful orgasms.

One woman I dated was the executive assistant for the president of a huge corporation. She was very proper and conservative, and wore Liz Claiborne suits. She owned a condo on a lake. I would take her out to the beach and fuck her on the picnic table where anyone with binoculars could see. I told her, "You love getting fucked where everyone can see you. You are so very naughty!" Then I brought her into her condo and fucked her in every room of her house telling her, "You love to get fucked! You can't get fucked enough!" She would almost lose consciousness with the long string of orgasms.

Women Love To Be Naughty!

I wanted to see how naughty I could make them be, even to the point of doing very taboo things.

Anal sex has always been a man's fantasy, but I thought women didn't want to have anything to do with it. How wrong I was.

I would make them feel very naughty, and then I'd use that to tell them that they are even naughtier because they want to be very taboo and get fucked up the ass.

Then I'd train them to have anal orgasms and then I'd fuck them up the ass! Very taboo.

Women Love To Be Taboo!

I had read stories about women being tied up, so I tried that.

I would tie her hands to the bedposts so that her arms were outstretched and she was restrained to the bed, unable to get away. Then I'd remind her how she is so very helpless to me and I intend to do with her as I please and have my way with her.

Then I'd tease her with oral stimulation and hold her right on the edge. Then I'd ravage her by fucking her hard, which always resulted in spectacular orgasms. They loved it!

Women Love To Be Dominated!

I pushed things further, even to the point of making her seem like a slut. I found that women responded powerfully to that.

I would say to women, while fucking them doggy style up the ass and pulling their hair, "You love to get fucked up the ass like the slut that you are!" They got even crazier!

As an extreme example of making a woman be a total slut, I brought one girlfriend to an on-premise swinger's club where sexual intercourse among couples happens in a large room as others watch.

I had her select a single man that she found attractive, and then he and I double penetrated her while everybody watched while I said to her, "You love to get fucked hard by two men! Show me how you love to get fucked! You love getting fucked hard like the slut that you are!" She was screaming in continuous simultaneous vaginal and anal orgasms.

Needless to say, she became even more wildly crazy about me. Powerful stuff.

It just goes to show...

Women Love To Be Slutty!

Now that does NOT mean that women want to BE sluts (in a derogatory way). Certainly NOT! However, in the correct context, with her trusted lover, women love to become ruthlessly expressive sexual creatures.

It appears that anything that is taboo is even more exciting, simply because it IS forbidden. It is that contrast, that dichotomy.

Simply the fact that it smacks at the general definition of what it means to be a "proper" woman is what makes it exciting.

Moreover, interestingly, the better the sex is, the more she has to have it!

You might think that when a woman is getting really good sex, she would be satisfied and not need it any more often. That could not be further from the truth. In actuality, when a woman is wildly crazy about her man and the sex is fantastic, she gets even hornier. She can't get it enough!

Women Are Horny Monsters!

In many cases, women are far hornier than men. One woman I knew who was in her late 30s told me, "I get so damn horny! I feel like a 16 year old boy. It's not fair!" Another woman told me, "It gets worse at the time I ovulate. Every man I see with a cute butt I want to jump him right then and there and ride him hard!"

Even so, women ONLY do these things when led by a man. They cannot do it themselves. They cannot initiate themselves. They can't even suggest it because that would be contrary to being a woman. She needs a man to lead her.

Women NEED Men!

Where is she going to find such a man? Men like this are extremely rare. That's what makes the Masterful Lover such a precious find.

Now let's get back to social conditioning.

Why did all that social conditioning want us to believe that women didn't like sex that much? It appears that there are a number of reasons for this.

Social conditioning protects women from men who only want to use women for sex. If social conditioning trains people to believe that women must first be in love to have sex, or must first be in a committed relationship, then women always have a defense against men who view women as simply objects. Rightly so.

It keeps young women in check. If parents teach their children that sex is only for love and marriage, then maybe their children won't go out and have sex and get pregnant. Certainly understandable.

Interestingly, the social conditioning does parallel the reality of being a woman to some degree. What I mean by that is the following. Most women are sexually submissive. They do not initiate sex. Instead, they respond to their man when he initiates. Most women prefer it that way. Women are excited by the fact that they excite their man. Women love to be desired and "taken" by their man. It makes her feel feminine and beautiful.

Unfortunately though, women are also victims to social conditioning. Social conditioning is interpreted to mean that any woman who loves sex must be a slut. This has caused great frustration for women.

Even though women love sex, they'd never admit it to a stranger. It would make them appear as a slut. It would also invite advances from men who objectify women.

The advantage of all this social conditioning is that since it is against social norms for a woman to love sex, it makes it even more exciting WHEN a woman demonstrates that she loves sex. Anything that is taboo is even MORE exciting because it IS taboo.

Always Leverage Taboos!

To further complicate things, most of us men have been raised to be proper polite gentlemen, and we have been taught that to suggest sex to a woman would insult her because it would insinuate that she is a slut.

That piece of social conditioning hinders men and puts men in a predicament. A man wants to be sexual with the woman he is interested in, but he can't suggest it because it would insult her.

The woman certainly can't suggest it because that would define her as a slut. Besides, women are sexually submissive. They prefer that the man initiate.

That's where YOU come in. You are going to be that precious find. You are going to lead her. You are going to allow her to finally experience ALL those things she has always wanted to experience but was too inhibited to do or had never met a man who could lead her.

Here is the most important thing that men need to understand, yet it is one of the toughest things for men to accept...

Women are far more sexual than men

In fact, women are far more sexual than most men can even comprehend.

Are Women Sexual Creatures?

Women are highly sexual creatures. Women crave really good sex, but for women, really good sex is rare to find.

Sometimes they have it, and don't realize how rare it is until later. Here is an e-mail from Barb:

I had my first boyfriend in college and he was an amazing lover, but I didn't realize how rare that was until later in my life, meeting other guys.

Once a woman has experienced it, nothing else will do. She will always crave it. Period.

Women will make the investment to make sex better, and they are very frustrated when their man is not willing to do the same. Here is an e-mail from Marian:

Hello David,

I came across your book through my brother. I feel like I am more curious and interested in finding out how to please my boyfriend more than he is. It seems like this book is for guys who are looking for more ways to be sexually creative with women, not the other way around, and I'm wondering if I will benefit from it.

To give you a brief understanding, my boyfriend is my first. He's 6 years older and sometimes thinks he knows everything.

I like going on the net and finding new things out about anything that interests me (sex being one of them). I don't have a lot of experience with it so I guess I like going and finding out how to please the guy in different ways, techniques, whatever. He just looks down on this kind of thing.

I would be so pleased if he was equally interested, so that it would become a joint venture that would bring our intimacy closer.

I don't expect him to change and I'm not trying to make him change. I just want to change and improve myself as a woman, a human being, a friend. I just want to be the best I can be.

Marian would really appreciate being with a Masterful Lover.

Here is an e-mail from Kayla. She is looking for a Masterful Lover:

Hi David!

I'm not completely sure if this is going to be a question you can answer, but after reading your e-mails I can't stop thinking about it.

I happen to fall into the category of women who have never had an orgasm, and knowing that there are guys (Masterful Lovers) out there that can make that happen is both encouraging and depressing. Why don't they find me?

How can I find a masterful lover?

I have met a variety of men. I went through the bad boy stage and now I'm looking more for respect and trust.

What I seem to find are men that I have to sort of "teach", in a way, how to please me (in which I don't even really know).

I guess I like the sexual intentions of the guys I meet because I like to be sexual more often than not, but they often lack the communication skills that I need and the ability to be in control. They always want me to take charge, but I won't do that until I am FULLY comfortable. They don't understand that.

Many women go through the bad boy phase, but after he turns out to be an inconsiderate jerk, she finally dumps him.

Still she will always be addicted to the excitement.

Eventually respect and trust become the important things on the shopping list.

As for having to teach a man, most women don't know how her man should please her, and no woman wants to have to be in the position to have to tell him. Many men don't know themselves, or they make assumptions, and don't care enough to find out.

It is no surprise that Kayla is a sexual creature. Most women are highly sexual creatures.

Further, women love it when her man can be in control, but many men cannot do it. In fact, some men are sexually submissive!

On occasions, a woman is willing to be in control, but only in the context of a long-term relationship built on mutual trust and respect.

Women like a confident man who can lead. Here is an e-mail from Cathy:

I appreciate the message that you are providing men, "Be the Man, be confident, lead her, etc." I wish that men could get their heads wrapped around the fact that they really

could have anything they wanted if they simply
knew the correct approach.

I couldn't have said it better myself. It really drives the point
home coming from a woman... "Men could really have anything they
wanted if they simply knew the correct approach."

Women wish men would study up. Here is an e-mail from Cindy:

David,

I am a woman who finally has discovered her
inner sexual being.

I have been getting your newsletters for
several months now and enjoy reading them.
They are geared mostly toward men and how they
can help their woman.

One problem I've ran across with my
"newfound" abilities, is that being a giver I
find men get "addicted" to a mind-blowing bj
and then forget about me. I have had men tell
me no woman has ever done to them what I have.
It is great to know you can please someone to
that extent.

Anyway, my question for you is: Is there any
way I can use your book to help myself achieve
things in a relationship where you know your
partner would never even consider reading the
book?

Your man will have to read the books himself. There is no other
way.

Sometimes women want their man to get rough. Here is an e-mail from Jenny:

Hi David,

I don't mind aggression when making love. Sometimes it really turns me on.

My man is an amazing lover. It's in the way he takes control of my body. He pulls and moves my body aggressively in different positions. It shows that he wants to give me sex that I enjoy. He would massage my vagina with his fingers very intensely. It's wonderful, even better than when he uses his penis.

I love it.

At the moment I can't seem to get enough. I just seem to want to do it all the time. I don't want to stop wanting it either.

May Jenny's e-mail serve as an example that women are highly sexual creatures. When it's really good, women can't get enough of it.

Sometimes, also, women actually like to be disciplined and restrained. Here is an e-mail from an anonymous woman:

Dear David

I was wondering what would be the best way to tell my boyfriend that I would like to be spanked while he is fucking me... or maybe even tied to the bed and "forced" to be submissive. I am not sure how to bring this up without him thinking that I am a freak.

Most women are sexually submissive. They actually like to be disciplined or be told what to do. This is normal and good.

However, being submissive causes a dichotomy. How can she tell her man what to do if she is by definition submissive?

Also, to do so would define her as a slut. Women do not want to be sluts, but they love to get slutty.

More on this later, but first, let's pose the classic question: "What do women want?"

What Women Secretly Desire

This discussion is in the context of women who are for the most part emotionally healthy. First of all, you should only be concerned about emotionally healthy women, and secondly, emotionally unhealthy women are driven by unhealthy needs.

This discussion is also in the context of what women look for in the men that they might consider for a relationship. This does not apply to one-night stands, as women do not take such men as seriously.

So, what do emotionally healthy women look for in the men they take seriously?

Don't Ask Women

When women are asked what they look for in men, they usually say something like "a sense of humor" or "a nice guy", but that is not what they really respond to.

The fact is that most women are unable to articulate what it is they really look for.

Every woman grew up knowing exactly the way it would make her feel to have the ideal man. It is that feeling she must have when a handsome stranger sweeps her off her feet. A woman does not choose a man because of who he is; she chooses him because of how it makes her feel.

What Do Women Want? It Is An Age Old Question

Certainly a woman who is 21 wants something completely different than a woman who is 31. We already know that women want a man that they can trust and respect.

So what is it? The trail of clues begins when she is very small...

While she was growing up and learning what it means to be female, she developed elaborate sexual fantasies. Early fantasies often begin as crushes. Typically she concentrated her romantic hopes and dreams on a movie star who she will never actually meet. In this context, she can imagine being in love without the complications of reality. It gives her mental practice at being desired, valued, and appreciated.

She grew up in the social conditioning that it is feminine and proper to be sexually submissive, and that it is unacceptable to be sexually forward. By being sexually submissive, she can enjoy sexuality without assuming blame for it. She can be "slutty" without being a "slut". Thus, she can be fully sexually expressive by simply following the lead of her man.

Therefore, in order for her to have a fulfilling sexual life, she is going to need a man who can lead her, such that she can experience those things that make her feel the way she has always wanted to feel. Thus, she needs a man that she respects, because she will only follow the lead of a man that she takes seriously.

In order for her to surrender to him, she needs to trust him. Only then can she give in to total abandonment.

She may also have fantasized about taming a rebel by teaching him how to love. This is the typical story in any romance novel. He is strong and independent. She makes him fall in love with her, a love affair that transcends all others. At the same time, she nurtures her own self-esteem and independence by imagining a lover who believes in her. He admires her for her intelligence and courage.

So, she needs a man who can lead her and she needs a man who will respect her.

Thus, the life of an adult woman is one of dichotomies. In her everyday life, she wants to feel like a lady, respected, and admired, but in the bedroom, she wants to be sexual; she wants to be fucked like the slut that she loves to be. That sounds like a contradiction, but she doesn't see it as that. It's what defines her as a woman.

The question really should be:

What Do Women Actually Respond To?

Reality is there for all of us to observe...

When you observe relationships, you find that some relationships are more fulfilling and successful than others. For the successful relationships, both people are getting all the things they need. In the relationships that are not fulfilling, one or more of either of the two people's needs are not being met. This gives us a clue as to what women need.

She Knows When Something Crucial Is Missing...

When you observe women who went outside of their marriage, you find that they are obtaining that which they were missing in their marriage. For example, if the sex is boring, she will get an exciting lover. If the sex in the marriage is acceptable, but she feels that she is not being treated with respect, she will find a man who will treat her with respect. If she does not have emotional intimacy in her marriage, she will go out and have an emotional affair. When women go outside of their marriage, they are looking to have one or more of their needs met. For some women, if it takes two different men in order to get all her needs met, she'll do it.

Examples Of Various Men In A Woman's Life...

The answer can be found by observing the various men that a woman has been with in her life. The various men met some of the needs, but not all of them.

The Bad Boy

For example, take the all-famous classic bad boy. The bad boy gives her exciting sex, and because he is very masculine, he makes her feel sexy. However, he is a self-absorbed jerk who cheats on her.

She does not feel appreciated, and she probably is unable to feel any emotional connection. She often becomes addicted to him because the sex is so good and she feels so very sexual. She will try in vain to tame him, but it is fruitless. In an emotional breakup, she finally leaves him for good. At least the high self-esteem woman finally leaves him for good.

You may be asking, "Are there lots of women that don't like the aggressive bad boy attitude at all? What about the important aspect of politeness and manners?"

A lot of women are turned off by the aggressive attitude of bad boys, if by "aggressive" you mean rude and pushy, and if the context is outside the bedroom in social situations.

Outside of the bedroom, and in social interactions, most women prefer "politeness and manners". It's just human nature.

Still, as you look around, you find that bad boys tend to get the girls, while nice guys don't. Women are often seen crying on the shoulder of a nice guy "friend" as she suffers through the extreme emotional highs and lows of a relationship with a bad boy "lover".

Bad boys are aggressive. Nice guys are polite. Thus, a lot of "reforming nice guys" who want to be "exciting bad boys" will try to act aggressive, but they just end up being rude wimps.

A lot of it has to do with context. As I have discussed at length before, women want to be treated like a lady in the living room and ravaged like the slut that they love to be in the bedroom.

Women know that bad boys can ravage them, and I'm talking about the dangerous to know, impossible to tame bad boy. (Also, I'm talking about MOST women, at least once in their lives.)

Bad boys are exciting. For a woman, the sex is awesome with a bad boy. He is dominant, he talks dirty to her, he makes her do things she is too inhibited to do, and she loves it. It makes her feel sexual. It is extremely exciting for a woman to experience these intense sexual feelings.

Bad boys are masculine. The feminine in a woman is attracted to the masculine in a man, and with a bad boy, it is extreme. It is this contrast that is so alluring to a woman. It makes her feel sexy and feminine.

Bad boys lead an exciting life. They are daring and live on the edge. Women want to be part of it to make their own life more exciting.

Bad boys are mysterious. They have a dark side that women are endlessly curious to know. She never really feels she knows him.

Bad boys cannot be tamed. Women are nurturing creatures, and are thus compelled to save him. She is determined to rescue him by teaching him how to love.

Bad boys are fearless with women. He will woo her by being sweet and gentlemanly, while she is drawn to his masculine confidence.

Bad boys are addictive. She becomes so wrapped up in her experience that she realizes she needs it. And the bad boy knows it.

Once she becomes addicted, the bad side of the bad boy comes out.

Bad boys are selfish. He inconsiderately takes and takes, and she gladly gives and gives even more in order to keep him because she is addicted to the exciting sex.

Bad boys are secretive. She begins to notice things he is keeping from her, things about his dark side that would be self-incriminating.

Bad boys make promises of fidelity and then deny their indiscretions. When she learns he has cheated, she is hurt, but she needs it, so she works even harder to save him.

Bad boys are paranoid. He knows that other men do exactly the same thing. Thus, he becomes very jealous and possessive of his women. He becomes controlling in order to keep her to himself.

It is incongruent. The internal incongruity is another form of weakness.

Most every highly sexual woman has at least once in her life been with a bad boy. This is true for both high self-esteem and low self-esteem women.

For the high self-esteem woman, she eventually can't take the drama any more, and she leaves him. It is probably the most emotionally traumatic thing she will ever do. It is heart wrenching for her because she has become so addicted to the awesome sex.

After a woman has been through the roller coaster ride with a bad boy, she will forever avoid inconsiderate jerks, but she will always crave those intense sexual feelings.

Let me repeat: She will ALWAYS crave those intense sexual feelings.

It is because she craves those intense sexual feelings that there is a tremendous opportunity for you. So you can thank the bad boys for doing some of the preparation work for you.

I have studied the bad boys for many years, because I wanted to have the same kind of powerful sexual affect on women, but I was determined to go beyond even that. I was going to figure out everything it takes to have that powerful affect on a woman and still have all the other things that a woman must have, because when all those other things are true...

It makes the sex even more awesome.

So when you step into a woman's life, she sees that she could have all the things she has always wanted, and she begins to become excited about the possibilities.

The Nice Guy

Then there's the classic nice guy wimp. He makes her feel appreciated, and he is able to establish an emotional connection with her. Unfortunately though, he's boring, or he's not masculine, so she doesn't feel sexy and doesn't feel turned on about sex. She'd love to marry him, because he'd be perfect for a family, but for some reason she keeps delaying the engagement. It's because he doesn't make her feel like a sexual creature.

Here is a letter from a woman ranting about her "nice guy" boyfriend. She posted it on craigslist.org. It is called "Why nice guys SUCK".

Best of craigslist - washington, DC - Why nice guys SUCK. Originally Posted: Sun, 10 Sep, 2006

Why nice guys SUCK

This is a long rant, so bear with me or hit your back button. I'm frustrated and in no mood for your shit either, so if you don't want to read it, well...

So I'm dating a nice guy now and it SUCKS. No other way to explain it, it just SUCKS. He's no challenge. He agrees with everything I say. He's got it all though - a decent job, a nice house, no kids, no psycho ex-wives, and he's tall and cute. Anyone ever seen that Friends episode when Alec Baldwin played Phoebe's boyfriend?

YEAH, my boyfriend is THAT nice. He's just too fucking nice. Nice is boring. I've never heard him raise his voice. He's never aggressive. He has no edge. He won't even drive over the speed limit and that fucking annoys the shit out of me, yet I sit in the passenger seat and keep my mouth shut... watching everyone whiz by us.

Don't get me started on the sex. Oh, excuse me... making love. After he cums (note I didn't mention anything about ME cumming), he rolls over and says, "Oh, that was nice," with a little sigh. I KID YOU NOT, he says it EVERY TIME and then he sighs like he has just woken from a refreshing nap. I finally got so tired of missionary and him looking lovingly into my eyes and smiling as he came, that I threw him down on the couch one night and mounted him. At first he was terrified - yes, TERRIFIED. He thought something had possessed me.

It HAD -- it was sheer MADNESS. I fucked the shit out of him that night. He then sighed and said, "Oh, that was nice."

Now that we had the cowgirl position conquered (always with that sigh afterwards), it was time to move on to doggie. His ex-girlfriend never did doggie. (Hmmm... maybe there's a "nice" ex-girlfriend to blame for his timid niceness? That bitch...) Anyway, I digress. I tell him I want him to fuck me from behind. Yes, I used the word "fuck" and I didn't care what he thought about it. He gets behind me and enters me, and damned if he didn't say, "OH, THIS IS NICE"!!! Are there any 35 y/o men out there that haven't smacked a woman's ass when doing her doggie? YES, and he's my boyfriend!

Tonight during sex, I think I'm gonna tell him to stick his finger in my ass when I'm riding him. THAT should be interesting.

So for the nice guys out there, my advice is this: It's great that you're nice (to an extent), but have some backbone. Don't be a spine donor all your life. When your girl is out of line, say something. Don't let her walk all over you. Occasionally, be a "bad" boy (being bad doesn't translate to abusive or criminal). Say "No" to her sometimes. Raise your voice and be heard. Say something dirty/sexy to her occasionally. Drink a few too many beers and piss out in public. Smack her ass.

Don't ever use the word NICE to describe things, especially sex (okay, that may be a personal pet peeve).

Have an interest in at LEAST one sport (or pretend to).

Drive 5-10 miles over the speed limit once in awhile.

```
Run an old lady off the road just for kicks
(yeah, I'm kidding about this one... just ride
her bumper for a few miles).

Be aggressive during sex. Take off those
damn white socks and Jesus sandals.

Grow a goatee for a few weeks. Shave your
balls. Stray from your routine and shake
things up.

BE A MAN FOR GOD'S SAKE... and the women
will fall at your feet.

Whew... THAT FELT NICE.
```

All too typical.

That woman is in the process of dumping her man, and she may not even realize it yet.

It happens often. The woman becomes frustrated that the man does not see the sexual creature in her. She still cares about him, but she no longer responds sexually to him. She thus loses respect for him.

Out of her frustration, she then goes into a stage where she mocks his ignorance by shocking him by making HIM do taboo things. After she has entertained herself for some time, she finds that she has lost all respect for him, and she dumps him.

The Emotionally Unavailable Guy

The closest she might get to the ideal relationship is the emotionally unavailable guy. He's not a wimp, and he's not a jerk. He may be masculine and give her good sex. He may be a gentleman and make her feel very appreciated. Still there's one important thing missing; he's not emotionally available. She never feels emotionally connected to him. This is probably the biggest complaint women have. She found the ideal guy in all respects, but damnit, he's not emotionally available. She tries over and over to get him to open up to her, but it never happens. Women are emotional creatures, and they need it, so she eventually leaves.

The Gay Friend

There are other men in her life. There's her gay friend. She loves hanging out with him because he makes her feel sexy. They have an emotional bond and he really makes her feel appreciated. She feels very close to him, but damnit, he's gay, and thus not an option.

The Mysterious Lover

And then there's the occasional mysterious lover. For many reasons she'd never ever consider him seriously for anything, but damn, the sex is good. Once every few months, when she is bored with how things are going for her, she'll call him up and go over for one thing, and one thing only: for the sex. Then she leaves hoping she will soon find the ideal man.

The Married Man

Finally, as one last example, there's the married man. Only low self-esteem women allow themselves to become a mistress, or in other words, a woman who wants what she can't have. He gives her exciting sex, he makes her feel sexy, and they share a deep emotional connection, but he won't leave his wife, so she doesn't feel special. Because she is low self-esteem, it is very difficult for her to leave. Let's get back to high self-esteem women.

So, those were examples of men in her life. As you can see, one or more of her needs are not being met in each example. By looking at what is missing, you can conclude what she has to have.

Women may not know it, and most of them certainly cannot articulate it, but women need four things in order to be fulfilled.

There Are Four Things Women Must Have

First, she needs to feel appreciated for the unique individual that she is. She needs to feel special, unlike any other woman, and she needs to know that her man supports her in her endeavors.

Second, she needs to feel that deep, intimate emotional connection. She needs to have that emotional intimacy with her man. It's a connection she shares only with him.

Third, she needs to feel like a woman. She needs to feel beautiful, sexy, and feminine. She needs to enjoy all those things that come with being a woman.

Finally, she needs hot, passionate sex. She needs to be seduced, enticed, teased, and satisfied, over and over again. She needs to experience new things, in new ways, including fantasies and roles. It makes her feel desired, affirmed, and alive.

If Only She Could Have All That She Needs

Imagine if she could meet a man who met all of her needs, a man who made her feel appreciated, a man who evokes a deep powerful emotional connection with her, a man who makes her feel feminine and womanly and sexy, a man who gives her hot passionate sex. It's the kind of thing women dream about.

Through her life experiences she begins to understand what she must have in a man. Some look for it in handsome men, some in popular men. She must have that feeling, but through a process of trial and error, including disappointment and despair, she begins to understand what she doesn't want in a man, like that guy who cheated on her, or that guy who was too much of a marshmallow.

If she meets a man who is exciting, she finds that he is reckless; so she looks for a man who is considerate, but he becomes boring; so she goes with a man who is macho, but he ends up being uninteresting; so she seeks a man who is intelligent, but he is sexually bland; so she takes a man who is dominant, but he is in every way demanding.

If she could only meet a man who had all the things that could make her feel what she needed to feel. A man who is exciting without being reckless, considerate without being boring, macho without being uninteresting, intelligent without being bland, and dominant without being demanding.

That sounds like a contradiction, but it's not. A man does not have to be reckless to be exciting, and he does not have to be boring to be considerate. Most men don't understand that, which is why women say....

A Good Man Is Hard To Find

So, when I talk to women, they light up when I talk about this. I will say to her, "I have asked women what they look for in a man, and what they tell me is..." (I hold out my left hand to the side.) "On the one hand, they find a man who is exciting, but he is reckless and inconsiderate." (I hold out my right hand to the side.) "And on the other hand, they find a man who is considerate, but he is boring." (Then I point both of my hands to myself.) "If there could be just one man who has both, who is exciting AND considerate." They love that, and it all gets linked to me!

It is obvious why it is important that all four things be there. A woman is most fulfilled when she is with a man that she has everything with, but here is the extremely important aspect of all four things being there:

When All Four Things Are There

Each Thing Becomes Even More Powerful

That is when she is with a man that she is wildly crazy about. That is when she can allow herself to become completely naked to him in every way. That is the pinnacle. That is when she is with a

Masterful Lover.

Good looks and love are not enough if the sex is lame. Here is an e-mail from Lisa:

David,

You are so right about the four things that a woman must have.

My last relationship was very loving and romantic. He was very good to me, and I loved him.

We could talk literally for hours. We would spontaneously dance in the living room. We were always touching each other. We would give each other "the look" and know exactly what we were communicating to each other.

The sex was always very loving and romantic… but the sex was ALWAYS very loving and romantic. There was never any variation whatsoever. We never talked in bed. It was always quiet and dark. It was never in the day.

Hence, I cheated on him all the time. One man, who was very exciting, gave me my first orgasm in intercourse. Then I'd go back to my boyfriend for all the other stuff. I never had orgasms in intercourse with him.

I am physically attracted to him. He works out five times a week. He just doesn't do it for me sexually.

Guilt got me. He is a good man and doesn't deserve to be treated like this, so I let him go. It was the hardest thing I ever did.

What is left is two people who love each other, who have a bond, but can't be together.

He now has a woman living with him. I get jealous that he is with her sexually. I know he just has her to have somebody, but I know she isn't good enough for him, and she is jealous of me, because she knows that he is still in love with me.

I miss him, but I know it would never work with him.

Here is a letter from Jean in Maryland:

Dear David,

I read your book "Secrets of Female Sexuality" last evening and I can't thank you enough for the valuable information. It just blew me away... finally someone understood me.

I am a very successful business women, and have many people (mostly men) working for me. I am in complete control, and I like that position, yet I have never understood why in other situations I don't want the control, I want to be led.

It finally became clear to me from the book that in an intimate relationship I want to be a WOMAN, and I want my partner to be the MAN. I don't want to make the dinner or date plans, I want him to, it shows that I am desirable, that I am a WOMAN (love that the most), and that he respects me. In a way I want to be taken care of, I want to be nurtured.

Also you said women need to be supported, so true, and we DO NOT want men to solve our problems, we really just need to emote, and have it accepted without judgment, just supported and admired. We can figure out the rest, we're pretty smart creatures.

I had all of this recently with a man that I

had started dating, and I was so surprised how sensual and sexual I was becoming / feeling. Our dirty talk was more via emails, and I saw it as "foreplay" and I was always ready for sex. I couldn't understand how I could have an orgasm with him just playing with my tits, always had found it boring before! Then he must have found either my G-spot or my deep spot, but almost instantly, my first real vaginal orgasm. Now I can just think about it without even touching myself I have an orgasm.

I was actually beginning to think I was becoming addicted to sex or to him, and it really scared me! I don't want to loose myself in a man, so I wasn't sure if I had a problem, if I needed therapy, maybe an intervention! Certainly this was not something I thought I could discuss with a girlfriend, what would they think, that yes I did need that intervention!

Recently we have had outside problems unrelated to our relationship, but both of us have somewhat pulled away from each other. Because of this I don't feel the appreciation I did, I don't feel the unconditional love (allowing me to emote, and him not trying to solve the problem), and the sex has completely changed. I no longer feel anything when he touches my breasts, I now see him as selfish in the bedroom, when before all I wanted to do was please him sexually. I tried to explain that our sex was beginning to make me feel like a whore, I didn't feel good about myself when I was with him. He was completely confused, how could he not be confused, and I couldn't really explain it to him, I didn't even get it myself at the time.

So I think it all goes back to the very beginning:

1. A Woman wants to be with a man who makes her feel good about herself when she is with

him.

2. We need to be respected, and yet feel desired, sexy, and always ready to be taken! (Oh god how I want to be taken!!!)

3. We need to feel as if we are being heard, and then usually we don't really have much we DO need to say... again respect and unconditional love / acceptance.

4. And most importantly, everyone needs to understand that Women want to be Women and treated like a Woman, and we want Men to be Men, and we will treat them like a MAN. A WOMAN wants to be both WOMAN, that responsible independent business women, and also that slut who craves sex anywhere, anytime, and in anyway that our MAN wants, or even better... directs it.

We women have had just as difficult a time understanding our roles in this every changing society as men have had accepting us becoming their equals. Yes, you are right that we want our men to be dominant in the bedroom, but also, to some degree, we want their domination in many other aspects of our relationship. We want doors opened for us, we want dates planned for us, we would love it if a man would plan the romantic vacation, instead of us, etc.

We can share equally the responsibilities on other issues such as what house to buy, how to raise the children etc., but please, let us be WOMEN and feel like WOMEN in our personal relationships. We will only, and CAN ONLY be a WOMAN if he is the MAN. Then he will start to understand the true magic of all of this, that as we feel more like a WOMAN he feels so much more as a MAN! What man wouldn't love that? It's a WIN WIN situation!

David, thank you again for the information in this enlightening book. I am so glad to finally understand myself, and know that I am truly normal!

My only question is: How do we delicately suggest this book to the man in our lives, without him feeling insecure about himself, wondering about his "performance" or lack there of? I truly believe this book is the key to a successful loving partnership. If everyone out there would read this book, all the therapists would go out of business; no one would need them anymore.

This should be a mandatory read for both Women and the Men we so dearly love and appreciate when they are being MEN and allowing us to be WOMEN!

Sincerely,

Jean from Maryland

The Dark Secret Of Women

As has been pointed out, women live a life of dichotomies. In fact, women need dichotomy in their lives in order to get everything they want.

The Dichotomies Of Women

She wants to be appreciated and proper, but she also wants to be ruthlessly sexual.

A woman wants to be known by the people in her life as being innocent, but with her man, she wants to be naughty.

During the day, she is that proper, educated, professional woman with a career, but at night, in the bedroom, she wants to express her sexuality like a slut. (Without being a slut. Big difference.)

During the day, she wants to be in control of her life and her destiny, but at night, in the bedroom, she wants to be submissive and told what she will do sexually.

Women don't want to be seen as sluts, but in the context of sexual intimacy with the man they trust, they love to be slutty.

In her everyday life, a woman does not like to be told what to do. But in the context of sexuality with the man she trusts, she responds very powerfully to being told exactly what to do.

During the week, she is very responsible, but on the weekends, with her man, in the context of sexuality, she wants to be totally without responsibility.

Women are repulsed when strange men send her pictures of their erect penis, but in the context of sexual intimacy, they find it a huge turn-on when their man gets an erection.

Women are repulsed when strange men masturbate while fantasizing about her, but she finds it very exciting when her man masturbates while fantasizing about her.

On Sunday mornings, she is a Sunday school teacher. On Saturday nights, she is a slut.

Women do not like it when men are possessive, relationship-wise, but in the context of sexuality, she loves to be "owned".

Women do not want to be objectified, but in the context of sexuality with her trusted lover, she loves to be treated like a piece of property.

Women are repulsed by the idea of being sexual with their father, but in the context of sexuality, she loves to call her lover "Daddy".

In The Context Of Heightened Sexual Arousal
EVERYTHING Changes

Certainly, every woman is different. Each woman has preferences, norms, and extremes unique to her.

The Dichotomies In Her Man

A woman is attracted to the dichotomy in her man. Her man must be sensitive, yet strong. Caring, yet dominant. Sweet, yet able to tell her what she will do sexually.

A woman is attracted to a man who is serious and focused about his career during the day, but at night can be light-hearted and people-oriented.

A woman is attracted to a man who is confident (not cocky – big difference) but who also does not take himself too seriously.

With the right man, a woman can allow herself to act dirty. (Without being dirty. Big difference.)

In the living room, a woman expects her man to use proper anatomical terms, like penis and vagina. In the bedroom, she expects him to use slang terms, like cock and pussy. If he uses slang in the living room, she sees him as inappropriate. If he uses proper terms in the bedroom, she just laughs.

When a woman feels secure, cared about, and special, then she can be turned on about being someone's horny bitch. She can be turned on about being fucked doggy up the ass while having her hair pulled and ass slapped. She can be turned on about being double penetrated by her man and his buddy.

All of the above assumes a mentally healthy woman with a good sense of deservedness. It is true that there are mentally screwed up women with a bad sense of deservedness, and practicing the above is more of an affirmation to her, rather than a healthy expression of her sexuality. (Very big difference.)

Another thing... Some men blame society, or religion, or strict upbringings, for repressing female sexual expressiveness. We know very well that women don't want to be sluts, but the reality is that they get slutty for us all the time. We hear women say they would never do a one-night stand, but the reality is that they do it anyway!

Society does not impose anything upon women, at least in Western cultures. Women choose it that way themselves. Women are social creatures. They want all that social establishments have to offer. If there weren't a societal structure, women would create one.

Women are social creatures. They prefer it that way. They want to know if the man is a social creature. They want to know if he can work within the social rules. They will run him through all their social tests, and when he passes, they fuck him silly. Simple really.

Women will continue to sustain the social norms and enjoy being on the arm of a proper gentleman, and women will continue to surrender to him when he fucks them hard like the very naughty sluts that they love to be. Stop worrying about society, or religion, or strict upbringings; instead, embrace the dichotomy of womankind.

Do Women Have All The Power?

Men complain that a woman can walk into any bar and take a man home in 5 minutes. From that, men conclude that women have all the power.

Even so, you don't see women doing that. What does that tell you?

Look at it from a woman's perspective. Imagine what it would be like to be a beautiful woman. You get offered free drinks. You don't have to put air in your tires.

However, it is because men want you just for your looks. You get objectified, and being objectified makes you other than who you are.

What if a beautiful woman did walk into any bar and take a man home in 5 minutes?

She would not know him. The sex would probably suck. She knows that she would just be an object, and then the guy would think of her as a slut.

There is no power in that because she wouldn't get all the things that a woman needs to have.

Women do not have any granted power to get what they want.

Look at the reality of the situation. The reality is that women are unfulfilled.

They read 400 page romance novels to vicariously live the life they really wanted to live.

They lay awake in bed at night fantasizing about what it would be like to be with a truly exciting lover.

Between ten and thirty percent of women have never had an orgasm.

Seventy percent of women have never had an orgasm in intercourse.

Reality points to the fact that women have all the frustrations.

A woman can only get all the things she has to have by being with a real man. A man who is exciting. A man who is a good lover. A man who will lead her into exciting experiences that she never dreamed possible.

A man who will make her do things she is too inhibited to do, but deep down inside really wants to do.

A man who's going to bring out that ruthlessly expressive, animalistic, natural creature in her. A man who she can totally surrender to, and be swept away with.

A woman wants to get slutty, but she does not want to BE a slut. She cannot initiate, because that would define her as a slut. She can only be slutty under the lead of her man.

Only with such a man can she have all that she needs to have, and thus, one could conclude that men have all the power.

However, the healthy way to look at it is that it Takes Two to Tango. You are the masculine compliment to her femininity. She is the feminine compliment to your masculinity.

Do not seek to have power over women. Instead, be Personally Powerful, and empower her.

Do not think about "getting sex". Instead, think about obliging her need for "really good sex".

Passion

Susan Shapiro Barash wrote a fascinating book called "A Passion for More", which sheds light on the dark side of female sexuality. It includes the interviews of 57 women who had extramarital affairs.

Every woman interviewed had something missing in her marriage. Each woman filled the missing need by taking a lover. Some of the women felt guilt about it, but most did not. In all cases, the women were happy they did it, as it ultimately led her to greater enlightenment and fulfillment.

For almost all of the women, they had much better sex with their lover than with their husband. For a few of the women, the sex was better with the husband, and the affair ended. In all cases, the woman's thoughts were consumed with the man that she had the most exciting sex with.

For some women, she had great sex with the man because she felt close to him. For most of the women, she became close to him because she had great sex with him.

I have taken a few quotes from "A Passion for More" to give you a flavor for the book. Because these are very minimal quotes, they are thus by definition taken out of context. Certainly, each woman's situation was more involved, but it gives you an idea...

Honey: "The affair awakened me sexually. I felt more uninhibited with my lover than with my husband. We had anal sex. He was a superb lover. I could have multiple orgasms with him, which wasn't happening in my marriage at the time. I was obsessed with him when he wasn't around, although we were together a few days a week."

Lucy: "The sex was excellent because he was so skilled. It was a learning experience for me. At first I slept with both my husband and this man; eventually I had sex only with my lover. It was hot. We had sex every possible way. For almost a year there was an intense, hot situation. The sex was wild and continual."

Sandy: "It was a wonderful sexual and emotional union and I never felt guilty. My lover adored me. He thought I was beautiful and gave me everything I wanted. The sex was loving and gentle but also uninhibited and open. We'd have sex everywhere. We did it in restaurants, in the back of the movie theatre. Wherever we were, we did it. If we went away, we'd do it on a mountain cliff, in the beach, in the car all the time, in the buggy ride in the city."

Hanna: "The sex was very exciting in the beginning. I'd see him daily sometimes, other times a few days a week, so there was lots of sex. Because he accepted me, I was able to play out sexual fantasies. The fantasies were things I would have done with my husband if he'd been willing, but with my lover there was no question that he was willing. My relationship with my lover made me see myself as a whole person and I rejected my husband's needs. My lover was reinforcing. He made me feel beautiful."

Rebecca: "Sex with my lover was very different than with my husband. Now I think of my lover as very handsome, but in the beginning my husband seemed handsomer."

In every one of the 57 cases, the married woman was not getting one or more of the four things that women must have. With her lover, she could get the thing or things that she was not getting with her husband.

I get e-mails from such women regularly. Here is one such e-mail:

My boss where I used to work, his name was David Shade, and all the guys used to snicker about his name and no one would tell me why, because I was the only woman attorney. One day I asked one of the senior partner attorneys, and he told me to check his name on the internet, and then I found your web site. Very interesting.

I have been married for 20 years. I have

been bored for the last 10 years, and I am ready to ask for a divorce. I met a man, and after less than 2 months with him I am crazy in love with him.

The sex is great. We are crazy in love with each other. We talk about everything and share everything we think and feel without hesitation. He is very sensitive and very much a man at the same time.

Both of us are very sexual and have connected to the things in the bedroom that turns us on, and we share them with each other. We have questioned the sex part over and over because it is the best we have ever had with anyone. All we know is that we turn each other on in ways we have never experienced.

Both of our marriages (both of us are married) are pretty much over and were really over before we met. Our marriages were over because of communication and sex not happening. So we have come to the conclusion that sex and communication now is the reason why we are so crazy in love, because we don't have it and have not had it for years with our spouses.

Well, this guy is the first one I have had and the first time I have cheated on my husband, and for some reason I do not feel guilty about it.

I bought your books and I am giving them to my lover.

He actually already understands 80% of this because he makes me feel like a woman that I have never felt in me before. I do not really think he needs these books, but this is really a gift to him to let him know he already has most all of it. This will give him even more confidence as a man in loving me and knowing that the woman in me really wants and needs all that is written in your books. I have felt it inside of me for a long time and this guy brings it all out. What a beautiful feeling it is.

I am learning about myself as well through them. Yes I am a naughty girl too :) but a professional during the day :)

Reading your material is giving me a window of understanding to know what I need to do differently for me to ultimately enjoy and have a very connected relationship with my new lover.

And here is a letter from a man who has benefited from the ignorance and neglect of married men. This letter appeared in Playboy Advisor:

I'm a single, average-looking businessman in my mid-forties.

During the past three years I've slept with every married woman I have desired.

I meet them in super markets, bookstores, and record shops.

I invite them for coffee, and the rest is easy.

From these encounters, I have observed the following:

1. I've not met a woman whose husband has made love to her properly in the past six months.

2. Many of these women had never had a multiple orgasm. Two had never had orgasms until we went to bed.

3. None of these women experienced any major

guilt from these encounters.

4. Most view they are neglected and view our time as luxurious sin.

In the mean time, I've collected a casual harem.

I'm never pushy; they call me.

Can you explain why so many men are such neglectful lovers?

Signed T.G.

That letter was not to illustrate that you should go out and pick up married women. Certainly not. It is simply to illustrate that there are a lot of men who are lousy, thoughtless lovers.

The sobering truth is that 30% of women have never had an orgasm. 70% of women have never had an orgasm in intercourse. Over half of all women have faked an orgasm.

Romance

While men look at porn, women read 400 page romance novels. Romance novels are the sexual entertainment for women. It is all about emotions and shared experiences. It gives them the chance to vicariously live the life they really wanted to live.

Nineteen leading romance novel authors contributed to a book called "Dangerous Men and Adventurous Women". It is a collection of essays describing the appeal of the romance novel. Edited by Jayne Ann Krentz and copyright 1992 by University of Pennsylvania Press, 186 pages. ISBN 0-8122-1411-0.

It is fascinating and very enlightening.

Romance novels account for half of all book sales. Because women love them so much, I was curious about two things: what do women find so fascinating in the men featured in the books, and what is the psyche of the women readers.

This book would lead me to believe that what women admire in themselves is courage. Both of these manifest themselves, or are played out, in her making that dangerous man fall in love with her.

The Plot: Victory for her is in teaching him to love. "Heroes who are gentled by love yet who lose none of their warrior qualities in the process, and heroines who conquer devils without sacrificing their femininity.

"It requires that the hero acknowledge the heroine's heroic qualities in both masculine and feminine terms. He must recognize and admire her sense of honor, courage, and determination as well as her traditionally female qualities of gentleness and compassion.

"It also requires a sexual bonding that transcends the physical, a bond that reader and writer know can never be broken."

The Romantic Hero: A warrior to be tamed. The romantic hero is a leader, strong, intelligent, untamed, tough, courageous, cynical, macho and dangerous. Yet, he is communicative, and deep down, is susceptible to succumbing to the vulnerabilities of deep love.

The Romantic Heroine: A woman of courage. The romantic heroine is a woman of intelligence, integrity, loyalty, faith, confidence, but above all courage. She conquers every challenge, the most difficult of which is taming the hero.

Sex in the Romance Novel: Sex is a celebration of the bond created between the heroine and the hero.

"He also happens to be wonderful in bed. It is a given that a woman is entitled to sexual satisfaction and that a real man can't be fulfilled unless his partner is also fulfilled."

The Female Reader's Psyche: Learning to love one's self. In reading the romance novel, the woman escapes into a fantasy world where she can be everything she has always wanted to be and feel everything she has always wanted to feel with the security of knowing that there will be a happy ending.

"If the romance novel teaches a woman to love anybody, the person she must learn to love is herself."

Fantasy

Women have fantasies far beyond the comprehension of most men. Female fantasies are often very involved and complicated, with entire story plot lines.

The classic reference on the subject is "My Secret Garden" by Nancy Friday, c 1973, ISBN: 0671742523.

Friday interviewed hundreds of women, and documented in intricate detail all the fantasies. It is fascinating and very eye opening. Most men will be surprised by how extremely sexual women are.

Another excellent resource is "Private Thoughts" by Wendy Maltz & Suzie Boss, c 1997, 2001, ISBN 1-57731-146-9, first printed in 2001.

This book is outstanding. It actually presents a thesis on the why and how of female fantasies. It explains many things about the female psyche.

The two main things I learned from the book are: 1. There is always a reason for a fantasy. Sometimes it is obvious, sometimes it is deeply hidden. 2. There is a lot of good self-talk out there and there is a lot of bad self-talk out there.

Maltz wrote that a woman's fantasy life is often a reflection of her search for sexual power, pleasure, and identity that started in childhood.

Maltz categorized scripted fantasies according to six different roles: Pretty Maiden, Victim, Wild Woman, Dominatrix, Beloved, and Voyeur.

Maltz wrote the following: "Sexual fantasies often begin in childhood with romantic or sensual ideas that become sexual as girls mature. Through fantasy, children create a very private, inner playground. It's a perfect place for girls to develop and explore all sorts of new ideas, including their private sexual thoughts."

"Fantasies help us by enhancing self esteem and attractiveness, increasing sexual interest and desire, facilitating orgasm, celebrating the present, satisfying curiosity, rehearsing future possibilities, releasing stress and tension, preserving a pleasant memory, and coping with past hurts."

Beginning as a little girl, a woman has been developing and exploring her sexuality, and what it means to be a woman.

She has been practicing how to use her beauty and charm to make a strong man fall in love with her.

Women grow up dreaming about being admired, about being seduced, and about surrendering to her man.

For women, sex is often about surrendering. She can only be ruthlessly expressive when she has no responsibility for the act.

When she is "taken" and "ravaged" she can be ruthless.

Taken to an extreme, it explains the popularity of the rape fantasy. No woman wants to be raped, but most women have masturbated to the fantasy of being raped, of being so desired, that a man would do anything to be with her, to ravage her, to fuck her hard. In this particular fantasy context, she can be as ruthless as possible.

The Rules Of Engagement

Some men have said, "On the first night, we get most of our clothes off, and I start fingering her. She gets off, then says, 'I've already gotten what I wanted tonight,' and I end up getting nothing."

That kind of thing happens on occasions with the young women if it is on the first night that you met her.

As for more mature women, they are more self-assured, interested in the man's pleasure, and decisive. If she goes home with a guy, it's because she has already made the decision that she is going to fuck him. If he gets her off before intercourse, she is even happier about her decision to fuck him.

As for mature high self-esteem women, they are especially interested in the man being excited about being with her. (This is in contrast to low self-esteem women who are looking for validation.) Also, they usually wait for the second or third date to let things go to intimacy.

Some mature high self-esteem women will only go as far as third base the first night of intimacy as a test of two things: 1) can she trust him, and 2) is she still in control. It is also a test to see if he respects her enough to wait until the next night to have intercourse.

On top of that, for the very discriminating highly sexual women, it is also a test to see if he can give awesome foreplay. These are the really fun girls!

In these situations (which you will find yourself in if you choose wisely) she is looking for a lover who is very good. If you have chosen well, awesome foreplay is in order. It will probably happen on the second or third date. Intercourse will happen on the next date.

Which brings us to the classic question: What are the long term ramifications of closing her the first night you met her versus the second or third date?

Let's look at two highly contrasting e-mails from two men:

Hi David,

I ordered your manual today and I think I have an important question. Your answer might be very interesting for at least some of your customers:

How am I able to avoid the girl/woman who wants a relationship?

I've been into a serious relationship for more than 12 years (I'm 32). Me and my girl broke up only a few months ago, so all I want to have at the moment is a good time. I'm successful with women, even the very beautiful ones - I think too successful...

I don't know what I'm doing "wrong". They all want a relationship with me. (I know this sounds weird since a lot of men out there would give their right arm for having my "problem".) I have not met a girl within the last few months (and there've been a lot) who would be satisfied by just having fun with me.

Since I'm not the type of guy who lies to girls to get what he wants there are only two things that happen again and again:

1. I know/feel what she really wants and I withdraw because that's something I don't want.

2. We do have "a good time" (at that point she already knows I do not want a relationship). The girl falls in love and there's drama afterwards. (I guess I better not read your book - it might be even worse afterwards. Lol.)

What can I do to avoid the girl/woman who wants more and still feel really satisfied? What can I do to prevent her from feeling like a slut by having a one-night stand or an affair with me?

Thanks for your help.

Mike

Mike, it is okay to not want a relationship, and you certainly are not in any condition to have one, since you recently ended a very long relationship. I hope you spent some time alone to get to know yourself again.

It is perfectly normal for a woman to want to have a relationship, and, certainly, her being in a relationship reduces her "slut" concern.

Nevertheless, there are plenty of women who don't want a relationship, but still want to have intimacy. They just want someone they are attracted to, that they feel comfortable with, and that they feel safe with. They don't need all the other stuff that comes with being in a "relationship".

This can be found with women who just got out of a long relationship themselves (and had taken sufficient time off to get to know themselves again).

Also this can be found among women who are very busy with some educational or career goal. They want to have fun, but don't want to be dedicated to a relationship. A lot of these women are high self-esteem women.

High self-esteem women who WANT a relationship are not going to waste their time with you. High self-esteem women who do NOT want a relationship would be happy to agree to your terms.

However, you're not attracting them (or you are, but you don't recognize it, or you give up too quickly).

That leaves low self-esteem women. Low self-esteem women would put themselves through the agony of trying to convert you, so that they can get validation. They don't think enough of themselves to avoid the drama.

Speaking of drama, you mentioned that term. It is a term

typically used to describe the lives of women who have low self-esteem.

If you are getting drama, you need to improve your screening skills.

Another possibility is that you are "doing" things that are in contradiction to what you "say". Women, with either high or low self-esteem, are going to go by your actions more than your words. You may "say" that you don't want a relationship, but the things that you "do" are viewed by the women as being congruent with starting a relationship.

Finally, yes, you absolutely must read my book when you receive it. It will give you a much clearer picture of all this.

Here is the other e-mail:

Hello there, Mr. Shade,

I've been reading a lot of your material. It's certainly helped me to have more fantastic one-night stands (many, many of these!). I still feel that I'm having trouble turning these into relationships though. I'm not sure if I'm going too far on the first night (i.e. offering everything, becoming a girl's dirty, domineering fantasy straight away) or, perhaps more likely, mishandling things over the following days (I normally text the girl within 48 hours).

I was just wondering what you thought about:

1) How far to go on the first night, and perhaps in more detail.

2) How to handle the follow-up if I am interested in repeating the night, with the possible view to turning it into a relationship (gradually of course).

Btw - it's not that I go explicitly puppy love on the girl straight away, far from it, but after a great night I can't help feeling enthused and texting sooner and perhaps more excitedly than I would otherwise, fearing that it will go dead if I don't.

Well, many thanks for your time. They should teach your stuff to guys in schools. Really, that would save us all so much time and would probably get us evolving faster....

Best regards,

Gary

Gary, most men eventually find a really exciting woman that they are very interested in and wish they could see more. Unfortunately, that outcome seems to elude some of those men.

In order to feel those really deep emotions that are really exciting, and in order to do the "really advanced" stuff, it requires an interaction that goes far beyond one night. So let's talk about how to make that happen.

First let me say two things: do not be in love with the idea of being in love. A lot of men have that problem.

What they end up doing is choosing a person who is not right for them, and certainly the relationship would be for all the wrong reasons, i.e. out of need. Being in a relationship is a reward. It is not a goal to seek. It is something that two people create together based on what they develop together as a result of who they are. To just seek a relationship is to take away from the genuineness of it.

Second, do not be afraid to be alone. It is better to be alone than with the wrong person. Believe enough in yourself to know that you will find what it is you require in a person. Have the self-respect enough to not stay with a person who is wrong for you, and have the self-esteem enough to be attractive to the really worthy women.

Now to answer your two specific questions:

1) How far to go on the first night?

If you want to see her again, do NOT go all the way on the first night!

2) How to handle the follow-up?

If you violate 1), then no matter what you do for follow up, you won't see her again. Usually.

From your e-mail, it sounds like you regularly violate number 1.

For some women, if she has a one-night stand, no matter how good he was in bed, she feels cheap. She will not see him again because she knows that he will never respect her as much as she'd like.

For other women, whether she specifically set out for a one-night fling or not, if she meets a man she finds attractive, and he pushes hard for the first night close, she just may take him up on it, for the fling of it, but that is all she sees him as. She has made the conscious decision to reduce the respect she has for him. He is not good enough for a relationship, he is only good enough for a one-night stand.

Now there are exceptions. I know a young couple who consummated their relationship the first night they met, and have been together for over a year and are very happy, but it is the exception.

Women Categorize Men

Women tend to put men into categories – one-night stand, relationship material, fuck buddy, just a friend, loser. There is no overlap. If she wants an anonymous one-night stand, she never sees him again. If she wants him for a relationship, she works on earning his respect.

Sometimes a woman can be moved from one category to another.

There are countless examples of successful relationships that started as friends and moved to lovers (but deeper inspection shows that most of those probably had some sexual tension all along).

It depends on the categories...

One-night stands rarely convert to relationships. Since a one-night stand is usually only about sex, she believes that the man views her only as a sex object, and she does not believe he could ever view her seriously. She believes that he could never respect her as a person.

Similarly, a fuck buddy arrangement rarely converts to a relationship. Usually, the fuck buddy arrangement eventually ends when one of the people moves to a relationship with someone else.

Interestingly, there have been a number of situations where I have converted a relationship to a fuck buddy, but I was unable to do it DURING the relationship, because that would violate the premise of a relationship. What I did was to tell her "let's just be friends". Then later, as friends, after some time had passed, we would reminisce about the great sex, and I would propose the fuck buddy idea. Since we were no longer in a relationship, and she already knew me to be an exciting lover, it was okay. In fact, if she didn't really want to look for a new relationship, but just wanted to have hot sex, it was a good thing.

There are rules for each category, and the woman follows the rules.

Women Follow the Rules of Each Category, and Only When it Does Not Violate Other Current Standing Rules

Don't be surprised if they DO stick to the rules! If you set yourself up as a once in a lifetime adventure, "It's Now or Never Baby", then don't be surprised if they take you up on it and then never come back! Even if the sex was good.

Why do women categorize some men into only one-night stands and others into only relationships? What makes her decide?

I thought it might be based on what she is looking for at the time. If she just wants a one night stand, that's what she's going to look for. If she wants a relationship, that's what she's going to look for. However, I began to find contradictions to that. Many woman have told me that they are looking for a relationship, but I end up closing them that night and never seeing them again. On other occasions women have appeared to be moving things along very quickly but then won't let their panties be taken off, and then the next day talking about introducing me to the family.

Often it appears that women don't know what they want, but I think most of them do. Ultimately, most of them would like to be in an exciting relationship with a very exciting person that they are completely fulfilled with and have everything they could have ever dreamed of. Sounds reasonable, but how often does that happen? What really is "ideal"? They have some learning to do. They need to have experiences.

They may meet an exciting man that does not meet all of their "relationship" criteria, but would be fun to fuck, so they have a one-night stand.

Sometimes they get into a relationship with a dependable person who might not be as exciting, but they want the companionship and regular sex.

Sometimes they get drunk at the bar and meet an ideal relationship material guy, but their inhibitions get displaced by their horniness, and they go for immediate gratification.

Basically, they fill the time. They learn on their journey.

Then there are those women who want what they can't have – they're called mistresses – and then there are those women who truly don't know what the hell they want – they're called flakes.

**Sometimes women take what works
for them at the moment**

I noticed among my friends that they were attracting certain kinds of women. Some of my buddies were always, and only, getting one-night stands, and others went from one long-term committed relationship to another. Could it be that women SAW them as being in certain categories, and went with that?

I began to notice that on the times that I drove hard for a first night close, it would most often end up that way, and not in a relationship. On those times that I took my time, if anything it ended up in a relationship.

Be careful what you wish for because...

Women go with the frame you present them

It all has to do with setting the proper frame. Decide what it is you want, and present that frame. Act in a way that is congruent with what you seek. If you want a one-night stand, set that frame. If you want a relationship, set that frame.

Some might suggest that you first find out what the woman wants, and then meld to that, but often they give the wrong information anyway. They may say they want a relationship, but only say that so they don't look like a slut when they fuck you that night. Some may move things along very quickly, but only to get into a position to sink their long-term relationship teeth in you. Most often though, the woman is waiting for you to tell her what you want, so she can decide if she wants to play by those rules.

The best thing to do is to define the rules and then let her decide. Yeah, you may miss one or two on the way who may happen to be looking for something specifically different, but being indecisive is placating and results in far more lost opportunities.

Of course, there are exceptions to everything, and of course the rules change somewhat as women get older and more mature and more secure in themselves.

The Secret To Giving Women Wild Screaming Orgasms

Females have the biological imperative to reproduce. Natural selection has weeded out the ones who are not driven to have sex. Women have the chemistry to be sexual, in fact horny.

The female orgasm serves two purposes. First, since it feels so good, she will want to do it often. Secondly, it serves to create muscle contractions that pull the sperm into the uterus.

Men have the biological imperative to reproduce. We are all very familiar with that. Men are also subconsciously driven to please women because it gives our sperm a better chance of making it to the egg. So it is completely understandable that we want our woman to have an orgasm.

Female sexuality is a mental thing though, and the female orgasm is very much a subconscious thing. She cannot consciously will it to happen. Similarly, if her subconscious objects to having an orgasm, for whatever reason, it will not happen.

Therefore, you must appeal to her mental sexuality.

While women have the chemistry to be sexual, it has absolutely no affect if the mental is not there.

Take for example the many recently divorced women that I had spoken with. In almost every case, she had become bored with her husband. Basically, she lost respect for him.

The interesting thing is she was no longer excited about sex. She may have even stopped having orgasms. She no longer could get aroused. She could not even get excited through fantasies. She didn't even masturbate anymore.

She would actually become convinced that there was something wrong with her.

Then she gets divorced. After some time she starts dating again and goes through a string of dorks and jerks.

Then... she meets an exciting man, a man who stimulates her mind, who leads her, who brings out in her that sexual creature that yearns to live. She becomes highly sexual and can't get it enough.

This illustrates how female sexuality is all mental. It completely overrides anything chemical.

Mental Orgasm

The most important lesson I learned from my phone sex period is that female sexuality is all mental. That is where the real power is.

I had become very good at giving women amazing phone sex using very descriptive narrative of what I would do if I were there. So I thought I'd push things even further.

I'd see if my voice alone could give her an orgasm. No stimulation whatsoever.

I would ask the woman which hand she was holding the phone with. Then I'd tell her to place her other hand behind her head and keep it there.

Then I would go into my usual narrative, but I would go into much more detail, excruciating detail.

I would start slowly and softly and sensually. That would go for about 5 minutes. Then I'd build it up a bit louder and a bit faster and more sexual for about 4 minutes. Then I'd go with very fast and loud vulgar description for about another 2 minutes.

Then they would have an orgasm!

An orgasm created by only the brain. No physical stimulation whatsoever.

Of course, they would be flabbergasted. They loved it.

This happened with many other women. It was always very consistent.

Since female sexuality is such a mental thing, I wanted the most powerful tool possible to leverage that, so I learned hypnosis.

Then things REALLY started to take off...

Since I could give a woman a mental orgasm, I wondered if I could give a woman an INSTANT orgasm, by simply telling her to come.

Instant Orgasm

After I made her come mentally, I'd let her rest for a while, and then I'd hypnotize her. This worked both over the phone and in person.

While she was under hypnosis, I'd simply tell her that when I say the words "come now" she will instantly have an orgasm.

Then I'd say, "Come NOW!" She would INSTANTLY have an orgasm!

After her orgasm subsided, I'd let her rest for only a few seconds, and then I'd say, "Come NOW!"

She would instantly have ANOTHER orgasm!

I would repeat the exercise enough times to make it permanently reliable.

Then, even after hypnosis, or even on another day, I'd fire it off as many times as I wanted! Fun stuff.

This has also been very consistent across women.

Extended Orgasm

The female orgasm has always fascinated me. It usually lasts for about 20 seconds. I wondered if I could extend it out further.

I learned from my phone sex period that I could extend a woman's orgasm out another 20 or 30 seconds by telling her right when she starts to come that I am about to come. Women love it when their man comes, and they want to come at the same time.

When she started coming I'd enthusiastically tell her that I am going to come. I'd tell her to keep coming because I am about to come.

After her orgasm had gone for about 20 seconds, I'd grunt deeply like I was coming and I'd tell her to keep coming. I'd make it very vocal and lasting for about 20 seconds. By doing this, I could get her orgasm to extend out to a minute.

I wondered if I could extend it out for as much as 5 minutes!

I just wouldn't be able to do it by continually telling her that I was about to come.

Since I had mastered the female orgasm through mental, I figured I could leverage that.

I'd cause a mental orgasm. Then with hypnosis I'd create an instant orgasm. Then another.

I would work her up to such an orgasmic frenzy that she was totally lost in it all.

Then I'd create one more instant orgasm and I'd simply TELL her that she was coming so hard that she couldn't even stop coming until I tell her to.

She would keep coming!

I would continue the intense vulgar narrative to keep her going.

This went on way beyond 5 minutes! It went on indefinitely!

I would get women to come for an entire HOUR!

Knowing what I learned from hypnosis, I decided to try doing it without any hypnosis. I got it to work!

I would not have to hypnotize a woman after all, and it was very consistent across women. They could all do it.

This further illustrates the fact that female sexuality is 100% mental.

Orgasms are the convincer. When she is having wild screaming orgasms beyond anything she has ever experienced, or even imagined, she is convinced that her man has a powerful sexual power over her.

Orgasms are addictive. A woman will become so addicted to the extended incredible feelings and the neurotransmitters, that she will HAVE to keep getting them! It causes her to become VERY bonded to her man.

Do it to your woman, and she will be CONVINCED. She will KNOW that you are her MASTER!

At least in the bedroom that is, which is the way she prefers it.

SECTION 2

How To Bring Out Her Inner Animal

One night, I watched her at the window in her sleep. I noticed for the first time, how a woman's underclothes barely touches her skin. How it rides on a cushion of air as she moves. How the silk floats about her body, brushing her flesh like an angel's wings, and I understood how a woman must be touched.
- Don Juan DeMarco

She taught me nothing new in matters of the act itself, but countless novelties in the way of sighs, ecstasies, transports and natural sentiments that arise only in such moments. Each discovery I made raised my soul to love, which in turn fortified me in the demonstration of my gratitude. She was astonished to find herself receptive to so much pleasure, for I showed her many things she had considered fictions. I did things to her that she did not feel she could ask me to do, and I taught her that the slightest constraint spoils the greatest pleasures. When the morning bells tolled, she raised her eyes to the Third Heaven like an idolatress thanking the Mother and Son for having so well rewarded the effort it had cost her to declare her passion to me.

- Giacomo Casanova

How To Bring Out Her Inner Animal

Again, this information only pertains to mentally healthy women. In fact, it best fits women who are intelligent, high self-esteem, and highly sexual.

When I say "women this" and "women that", I mean MOST women. Certainly there are exceptions to everything.

Also, this only applies to women who you intend to see on a continuing basis. It does not apply to one-night stands. The most important reason for this is that only women who take you seriously are going to respond to this.

Remember, when you lead a woman, you are responsible for her. You must be a responsible leader.

Further, all of this is absolutely in the context of enhancing a relationship that is based on trust and intimacy. It is all for the enhancement of the relationship.

Basically, this stuff only works on "the really worthy ones", as it should.

This is not about having power over women. This is about "being powerful" and about "empowering her".

Most importantly, you want to get your woman to a point of total abandonment.

Always know that women actually like sex. A lot. They like it often, and they like it good. If a woman does not want sex often, it is because it is not good. Follow this general rule:

The Better It Is, The More She Has To Have It

Once they do have it, they become addicted to it. Nothing else ever compares. You can ruin it for all the other guys.

If the woman has been with a truly exciting lover before, that's actually a good thing, because she has to have it again. She will always long for it. Once she meets you, and she begins to recognize that you can make her feel all those things again, she will powerfully respond.

Unleash Her Sexual Potential

Men need to understand that there is a massive amount of untapped sexual potential in every woman. It is there in a woman. Everything you need to fulfill a woman is already within her. You simply need to tap into that. Only a man can do that for her. Thus, men have a tremendous amount of power. More than they know.

Many men complain that women have all the power. They argue that a woman can simply walk into a nightclub and say to a man, "Let's go and have sex," and it's a done deal. But those men do not understand where the real power is.

What she really wants is a personally powerful man who will lead her through all those experiences that bring out the natural woman in her. Women may not know this, and they certainly are unable to articulate it, but it is exactly what they respond to. Such a man, in turn, is an extremely rare thing to find. Such a man has a tremendous amount of power, because she has to have that, and she reacts so very powerfully to it.

Feel Like A Man

Ultimately, all of us men want to feel like a man. We have various ways of experiencing that. For some of us we look for it in sexual conquests. For some of us it is conquests of power or money. For most men, however, I have found that he really feels like a man when his woman believes in him completely and does all those naughty taboo things with him.

Napoleon Hill wrote that "man's greatest motivating force is to please woman".

If a man is with a woman he can't please, he will go on to the next one, and women need to understand that it is very important to him to please her. If she doesn't give him that opportunity, he will go find a woman who will.

Basis of All Interaction Between A Man And A Woman

The underlying basis of all social interaction between a man and a woman who just met is sexual, plain and simple. It is not articulated, it happens subconsciously. It is what is on both people's minds.

You are the masculine complement to her femininity. She is the feminine complement to your masculinity. There is nothing more natural and normal than a man and a woman getting together.

The Correct Motivation And Intent

Don't think about getting sex. Instead, think about giving her really good sex. There is certainly opportunity out there. That is the correct motivation.

The correct intent is to bring out the decadent, primal, carnal, insatiably hungry, ruthlessly expressive, natural sexual woman in her. That will drive all the rest.

Genuinely LIKE Women

Many guys see women as the obstacle to get around to get to the sex. Wrong!

Many men are angry with women. They blame all women for the few women who dumped them or shunned them. Wrong! There is not a grand scheme among women against you. Quite the contrary. Women are actually hoping for men. Women honestly and truly want men to fulfill women. However, women usually cannot articulate how men can fulfill women.

Instead, believe in women. Have empathy and compassion for women. Be fascinated by women. Genuinely get to know and understand women. The more you do, the more fun it becomes.

Command Respect

Demand that you be treated with respect. Treat others with respect. Associate only with those who treat you with respect. Treat yourself with respect. A woman can only have respect for you if you treat yourself with respect.

To command respect, you are self-reliant and a responsible person.

You must command respect in order for her to take what you say seriously. She must take what you say seriously in order for her to become responsive to you.

At the same time respect the fact that she wants to be slutty!

Earn Her Trust

In order for her to surrender to you completely, she must trust you completely. You must earn her trust. You do this by demonstrating that you are trustworthy.

This is very simple. Do not violate her trust. Do what you say you are going to do. Do not do what you say you are not going to do. Rise to the occasions in all the little opportunities that make themselves available for you to demonstrate that you are trustworthy.

Be A Man

Stand tall, tell the truth, live a life of integrity, and live up to your responsibilities.

Many men are afraid to be a man. As you look back, you can probably find that most of your bad decisions and failures can be directly attributed to the fact that you were afraid to be a man. I know I can certainly say that.

Those men who get told by women "let's just be friends" are men who are afraid to be men. They are afraid to assert their masculinity in order to not risk the friendship they have with the woman. They think that if they voice their attraction for her, that she will dismiss them. In some cases it may happen, but certainly not in most. Such men are left in a drought.

Passion And Absolute Conviction

All of the men who are successful at bringing out the natural sexual creature in their woman are men who are passionate about it.

You must do this with absolute conviction, absolute determination. There can be no crack in your armor, there can be no crack in your voice, or it won't be believable. Only when it is believable is it going to work. You are leading, and she will only follow if she sees you as believable.

Have A Plan

When you bring your woman out on a date, know what you are going to do. Know what time it is going to happen. Know where you are going to take her for dinner. Know where you are going to take her dancing.

Of course, you are flexible too. If she suggests something else, you can do that.

In our modern world, women are busy during the week with their lives, such as working, getting an education, and so forth. Women want to be respected for their intellect and accomplishments, as they should. That's all good. On the other hand, on Saturday night, she doesn't want to have the responsibilities of the date. She wants to feel like a woman. She wants to enjoy the classic role of being a woman. She wants the man to be the man.

That sets the basis for everything. It is a context in which the man leads and the woman follows.

Romance

Romance is the socially acceptable term for sexuality. In the context of romance, sexuality is acceptable.

There is a lot of symbolism in romance. Take advantage of those symbols. Learn how to fold an ordinary paper napkin into a paper rose.

In the context of romance, she can begin to feel all those things that women want to feel. It's just like the romance novel.

You don't have to be in love to have romance, and you don't have to be in a classic dating situation to have romance. Also, being romantic does not mean that you are in a classic dating situation.

Romance elicits all the wonderful dreams she had as a little girl, and that is the context in which sexuality can emerge.

Emotional Connection

Emotional connection is the lubricant to sexuality.

Women are emotional creatures, and they want to feel connected to their man.

Establish that emotional connection. Allow her to emote, but don't let it go so far such that you end up being her emotional pincushion.

Do not try to fix her problems. Women loathe that. Instead, have empathy for her and support her. Women love that.

Show your vulnerability to her. Allow yourself to emote to her, slightly, not in a needy way, but in a way to convey connection and to slightly reveal your vulnerabilities.

How can you be an emotionally connected person but still be the man and command respect? Being emotionally connected is a manly thing. It is not contradictory.

It is only a wimpy thing when it is with lack of sexual threat. It is only wimpy when the man is sexually harmless.

Calibration

The truly skillful lover simply knows when his woman is ready. You should know when she is ready to be kissed. You should know when she is ready to be entered.

There are many tests out there that tell you how to know if she is ready to be kissed. You can run them, but you need to get to the point where you simply know.

The problem with running tests is that you are not as believable. You no longer come across as being in the lead. You are not as much of a sexual threat if you are testing. Instead, you should be calibrating and leading.

You will if you get out of your own head. Pay attention to her. Enjoy the process. Enjoy watching the changes in her.

The "It Just Happened" Clause

Everything you do must be in the context that she can simply say, "It just happened."

Do not articulate what is going to happen. Instead, simply lead her.

What is going to happen should never be articulated, or else for her to comply is for her to articulate what is going to happen, and to do so would define her as easy, or a slut, or being in the lead. That is not the woman's role.

Instead, simply move things along slowly but surely using body language. Let her get lost in the passion.

She should always be able to justify it later by simply saying, "It just happened." Monday morning she can go into work and tell her girlfriends all about it, and then say, "It just happened," and they'll all swoon, because they know exactly what she's talking about; two people got swept away in passion.

There will be times later in the relationship when you can, and should, articulate what is going to happen, for effect.

Patience

Too many men are in a big hurry to get to the sex. That is a huge mistake.

Do not push against last minute resistance. You will be like all the other guys who objectify women and do not respect her as a real person, and it will reek of desperation.

Instead, defer your own gratification for the big goal. The big goal is to bring out that ruthlessly expressive sexual creature in her. It's going to take a little time before she becomes your very naughty horny little cum slut fuck bitch.

In fact, move things along just slightly slower than she'd like it. Make her wait. It builds that sexual tension, and it makes her think. When she is away from you, she is going to think about it a lot, because it is that time away from you that she becomes close to you.

That is early in the relationship. Later in the relationship, there will be times when you will initiate sex when she least expects it.

Dominant

Most women are sexually submissive creatures. They respond to a dominant man. Be comfortable in that role.

Here's a good example. If she is wearing a T-shirt or a sweatshirt, pull the shirt over her head, but keep it on her arms. This lightly restrains her arms. It's called "The Hockey shirt trick".

While laying down and making out with her, hold her hands above her head.

While standing and making out with her, hold one hand behind her head, and the other behind the small of her back. Then move her backwards until your elbows smash into the wall. She will not be hurt; your elbows will take all the blow.

It is all symbolism. Many of these light restraints she could easily get out of, but it's the symbolism that is powerful.

The Remote Control Egg

This is an egg shaped vibrator that is turned on and off by radio remote control. You hold the remote! A totally awesome way to demonstrate your dominance. Makes dining out most interesting.

You can read a story on my blog of how I have used this: DavidShadesBlog.com

Show Her How Much She Turns You On

This is the magic dust of instant turn-on for her.

Tell her how much she turns you on. Demonstrate to her how much she turns you on. Tell her how you think of her when you masturbate.

During intercourse, especially at the climax, be expressive. Women love to know that their man is excited about them.

Do all these things. It works like magic.

The Importance Of Naughty

Naughty is the enabling context for taboos.

Everything taboo that you bring her through will be done in the context of naughty. She's doing it because it's naughty, and that will get her through all those taboo things that she would never do.

When she's with you, because you drive like a man, you are dominant; simply tell her that she is being naughty. In that context, nothing is taboo.

Also, be sure to tell her that it turns you on! That will be her justification.

Dirty Talk

I cannot tell you how very important and how powerful this is. Your most powerful tool in the bedroom is not your tool; it is your voice. It is by far the most powerful tool.

A lot of men have trouble doing that. They are concerned that she will take it wrong, or that she will be offended or insulted, or that it would make her a slut.

Women don't want to be a slut, but they love to be slutty. In the right context, with her man that she trusts and respects, in privacy, under his care and direction and watchful eye, she loves to be slutty.

Start things slowly. Build the sexual arousal. Then begin introducing the dirty talk. Let the sexual arousal increase. Be sure to do it with absolute conviction. It must be believable.

As the sexual arousal increases, the context changes. Everything you say in heightened sexual arousal is taken in a completely different context. She will not take it wrong, because:

In The Context Of Heightened Sexual Arousal

Everything Is Taken

In A Completely Different Context

It is all mental stimulation. I explain this in detail, including an interview giving a woman's opinion, in my "Sexy Dirty Talk" CDs: SexyDirtyTalk.com

Phone Sex

Phone sex is dirty talk on steroids.

In phone sex, you can get very raunchy, and she can respond powerfully to you. You start off easy and then slowly ramp it up to be extremely explicit and crude and rude.

You can get away with things in phone sex that you have not yet been able to get away with in person. Someday they may be possible.

In phone sex, you bring her through experiences mentally. Once she has experienced it in her imagination, once she has made it real in her mind, and it is associated with heightened sexual arousal, she can later experience it in person.

Because it is phone sex, it is just a fantasy, and so she can go along with it, but once she experiences it in her mind, it has become real for her subconscious, and thus, it can later become true.

For excerpts of actual phone sex recordings, see: GiveWomenHotPhoneSex.com

New Orgasms In New Ways

Orgasms are the convincer! Without wild screaming orgasms, nothing will work.

Give her new orgasms in new ways.

If she's never had an orgasm, give her her first orgasm. If she's never had a vaginal orgasm, give her a vaginal orgasm. If she's never had an anal orgasm, give her an anal orgasm. If she's never had a nipple orgasm, give her a nipple orgasm. If she's never had an instant orgasm on command, give her an instant orgasm on command.

Bring her through these new experiences. They are powerful convincers that increase her responsiveness to you.

It's not just that you give her orgasms, it's that you have a lot to say about when and how long she has an orgasm. Be dominant!

The "Yes" ladder

Lawyers have a saying, "A lawyer never asks a question in court that he does not already know the answer to." I would add to that,

"A true lover never asks a question in bed that he does not already know the answer to be 'Yes.'"

Get her sequentially saying yes.

"You love it when I do this." "Yes."

"It feels so good when I do this." "Yes."

"This makes you so excited, doesn't it?" "Yes."

"You want me to ____." "Yes."

This sets up a pattern for her following your lead.

Fantasies And Role-Play

Get to know her fantasies. Describe them back to her as you masturbate her.

Make up some fantasies. Lead her mentally through those fantasies as you masturbate her.

Introduce new fantasies using phone sex.

Use role-play to act out fantasies. Since it's just "make-believe" she can feel much more comfortable about playing the part.

Once she has played the part, it has become real for her in her mind, and she is much more apt to actually do it for real.

Expand Her Sexual Experiences

With you, she's going to do all those things she dreamed about experiencing, and even some she never dreamed about.

You are going to lead her through these experiences, such that you and her are going to experience these things together in real life.

You are always pushing the envelope just a little bit. Little baby steps at a time.

All of these things that you bring her through continue to open her up, continue to release that massive sexual potential.

The Power Of Possessive

It is a bad thing to be possessive of your woman from a relationship standpoint. It demonstrates fear and weakness.

However, in the context of heightened sexual arousal, it is a very powerful thing to be possessive of your woman.

"Who's naughty little slut are you?" "Yours, Daddy."

Piece Of Property

For the really advanced level (brace yourself) treat her like a piece of property.

Bring her to a fetish party and pull her around by a collar and leash.

Make her wear lingerie and heels around the house as you ignore her.

After doing this for an appropriate amount of time, you then take her and use her to get off.

Apply with a thick coating of dirty talk. Not for the faint at heart.

Of course, all of this is done in the context of equality in all other areas, and this is done in the context of absolute respect. In the context of absolute respect for her, she is secure in knowing that she is your special woman, you are ultimately doing this for her sexual fulfillment, and she can totally let go and totally surrender to the passion. In this context, she can assume the 'role' of being your piece of property, which is the ultimate expression of sexuality for a sexually submissive woman.

Total Surrender

Get her to the point of total surrender, making her completely naked to you in every way: physically, emotionally, and spiritually.

Post Coital Re-Bonding

After an emotionally powerful sexual experience, you MUST ground her again. You must make her feel close to you and secure with you. You must demonstrate to her how very important she is to you and how much you appreciate her. You must re-establish that emotional connection and intimacy.

This prevents her from later feeling sleazy and cheap, it makes her become even closer to you, and it makes her look forward to the next episode even more.

To Be A Masterful Lover

You love women. You are fascinated by women. You adore everything that defines a woman as woman. You derive no greater pleasure than giving a woman the most powerful, long-lasting pleasure of her entire life.

A woman is most fulfilled when she is wildly crazy about her man; when she feels beautiful and feminine and sexy; when she feels naughty; when she is aroused beyond description; when she can't get enough.

You appreciate the massive pent-up sexual potential in a woman, and you bring that out in her, until she surrenders herself completely to you, allowing herself to become totally naked to you in every way.

A woman's pleasure is very cerebral. Her mind is the most important factor. You drive her wild with desire.

You understand that women are highly sexual creatures that crave intimacy, and you are comfortable with her sexuality. Women are far more sexual than men. Women have elaborate intricate fantasies beyond anything men can imagine. For women, it is largely mental and emotional, and about being swept away.

Being a Masterful Lover is about being personally and sensually powerful.

Being personally powerful is about empowering beliefs, self-reliance, a sense of deservedness, and Being The Man.

Being sensually powerful is about enabling her to enjoy everything that comes with being a woman, by expanding her envelop of sexual experiences, and bringing out in her that ruthlessly expressive natural woman.

Everything you need to completely fulfill a woman is already within her. You bring it all out in her because you command respect, you earn trust, and you lead her.

Too many men see women as an obstacle to get around to get to the sex. Such men are just masturbating inside an object.

What you are seeking so relentlessly is nothing that comes from outside of you; it's something you already have.

The key to having what you want is not getting what you want, but being the person for whom getting what you want is a mere byproduct of the reality you create simply by the way you live life.

It builds a foundation of belief that is very real and a relationship with reality that is very empowering.

It is about two people celebrating being human beings. She is the feminine complement to your masculinity. You are the masculine complement to her femininity.

She knows that you can make her think powerful thoughts and feel powerful emotions. She becomes ultra-responsive to you.

There is nothing more natural than a man and a woman being together.

Stop thinking about getting sex from women. Think about giving women really good sex. Women are highly sexual creatures, and you're the man who's going to oblige her needs!

You are the man who's going to leverage her emotional soul to connect with her, her romantic heart to entice her, her innate sensuality to excite her, and her downright horniness to satiate her, and then do it again, and then again, until she is delirious with pleasure, ecstatic about being a woman, totally loving life, and wildly crazy about you!

SECTION 3

Success Stories

Have you ever loved a woman until milk leaped from her as though she had just given birth to love itself and now must feed it or burst? Have you ever tasted a woman until she could believe that she could be satisfied only by consuming the tongue that had devoured her? Have you ever loved a woman so completely that the sound of your voice in her ears would cause her body to shudder and explode in such intense pleasure that only weeping could bring her full release?

- Don Juan DeMarco

Be it real or feigned, her sleep tells the intelligent lover of an adored object that he is unworthy to enjoy her the moment he begins to wonder if he may or may not take advantage of it.
- Giacomo Casanova

Success Stories

Here are a number of e-mails that I have received from clients. Each is an example of how the man had applied the teachings to empower his woman.

The first e-mail comes from Dave in Canada:

I recently bought and read "The Secrets Of Female Sexuality" and "Give women Wild Screaming Orgasms". I must say it's changed me and taught me a lot.

Before getting your books, I had acquainted myself with one of the most attractive women I have ever met. She's talented in arts, successful professionally, an accomplished and elite athlete, and a complete knockout. The guys at work nearly always talk about how hot she is.

I became interested in her and she reciprocated. We hung out a couple of times and were both attracted to each other and began having sex. She mentioned often that I was the best she ever had, but she had never had an orgasm, and thought that something was wrong with her.

As our relationship progressed, I became increasingly paranoid that she couldn't like me. I constantly wondered when she was going to call me and call the whole thing off. (She did at one point, but came back later and I of course just took her back immediately, no questions asked).

When we were together I was always afraid of leading her, making the first move, being assertive, and just being a man in general. I was constantly indecisive, always wondering if she was happy, never suggesting anything specific because I was so damned intimidated. I started trying to prove myself to her by always saying things that I thought would impress her, or make myself seem worthy.

Then she started saying she'd call me later and wouldn't. She'd show up sometimes 45 minutes late for a date and didn't seem to care. I was always "okay" with that, though it made me feel shitty.

We agreed to see a movie one day, and despite my constant txting her, she never responded or showed up. When we saw each other at work the following week I pretended like it was no big deal... I think you get the picture.

Then I read your books and realized that I wasn't being a man. My subconscious was sabotaging the entire relationship because it deemed myself unworthy of her.

I started remapping my subconscious world through some techniques I learned in a meditation class (I began to see my entire world differently) and in a week I knew what I had to do.

It was important for me to take responsibility for my actions and not be the victim. Though I wanted so badly to make her feel bad, blame her, and label her a disrespectful person, I admitted to myself that I wasn't getting respect because I firmly believed that I didn't deserve it. By taking responsibility I felt empowered to change my situation. No person is worth sacrificing your dignity, no matter how hot and perfect they are.

So I called her and said, in a nutshell, the whole thing wasn't working for me, that she's a really great person but that I needed to walk away from it because I wasn't being respected and I refuse to be in a situation like that. When she got defensive, I assured her that she's a good person, I held nothing against her and didn't blame her one bit for what happened. I felt liberated and strong.

That afternoon she asked me to meet her to chat for drinks, which I accepted. David, she started looking at me differently. Her eyes sparkled and she hung on my every word. She was really open and expressive to me. We decided to get together the next day, and she was constantly keeping me up to date on where she was at, and whether she was running a bit behind.

That evening she came over and had her first orgasm (deep spot method). Then I lead her to another. We ran out of time to go for a third, but I felt so empowered that given the time, and evening of constant orgasms seemed imminent.

As I held her after the second, she laid there staring at me and gazing at the ceiling in disbelief (I was amazed too). I gave this perfect 10 hottie an unforgettable experience that no one could do before me. I not only feel lucky, but that she's lucky to have a strong man unafraid to lead her in a proper way.

I understand what this woman (and any other woman) wants and needs. I'm willing to walk away from any situation that isn't in line with my self-concept, including this woman who I totally adore.

I want to reiterate how important self-esteem is. If you don't respect yourself, if you're desperate, afraid of being a man with confidence, nothing will work.

You can't have the unassuming, strong yet not overbearing confidence that women desire without an unwavering belief in yourself as a person that deserves all the right things. This was a powerful lesson for me David.

Thank you.

Here is an e-mail from Robert in Cambridge, MA:

I have your "David Shade's Manual" and I've found it very helpful, as are these newsletters that you send out. The first step is always realizing that something is possible, and then all kinds of amazing things happen.

I am currently seeing a woman who has been blessed with some phenomenal gifts, and just KNOWING that things like extended orgasms are possible (and listening to your interview and reading your book and following your advice) has allowed me to give her orgasms that have lasted close to 10 minutes (clitoral with oral and digital stimulation). Your latest newsletter encouraged me to try some post-coital suggestion, which has done wonders as well.

Nothing breeds success like success. So now, after she has achieved incredible pleasure in ways that she never thought possible, I have convinced her that everything is possible and that it WILL happen. She has now started to experience extended vaginal orgasms as well. She now routinely has 5 or 6 orgasms per session, she can have orgasms clitorally and vaginally, and, suffice it to say, she is a very happy girl.

One of the most powerful things I've gotten from the whole experience is that women want to be led and if 'I' believe something can happen, then I can make it her reality as well

-- and truly, I learned that from you. Hopefully there is something in my e-mail that might let others believe that it is not something that only David Shade can make happen.

The other incredibly powerful thing I got from you is the idea that you have to know how to ravage a woman -- I think it's in your primer about what women want.

David, this girl knows that I am simply going to take her whenever and wherever I choose -- and she loves it.

Now, in the past, I would have been way too timid to ever think I could get away with something like that -- thinking that I might upset her, or seem too much of a jerk, that I might only want her for sex, etc, etc, etc.

With this girl, one night early in our relationship after I'd given her the first extended orgasm she'd ever had (it was maybe 30 seconds at the time), I told her, "Look, we seem to be getting along really well, so, just to be honest with you, I want you to realize this from the beginning, but I am going to ravish you constantly. I am going to fuck you as often as I want and wherever I want, and I'm not going to stop. I'll know when you REALLY don't want to, but in general, I'm going to just take you as I please. This is the way I am, and I'm not going to change, so if you think that's going to bother you, I want to give you the option now to leave."

Of course, I half expected her to get upset or to equivocate or somehow try and take some of that power back. Instead, she broke into the biggest smile and literally screamed, "Baby, do you really mean it! I mean, really, you're not just saying that?" Of course, I truly spoke from the heart and told her that was the way that I am -- and when I wanted her, I was simply going to take her. She was so happy.

So, of course, I have. David, I literally fuck this girl every single time I want. She'll be cooking breakfast and she'll know I'm looking at her ass -- because I'll tell her flat out. Then I'll say something like "you'll be lucky if you get to finish that omelet before I fuck you". She'll giggle and get a little scared, but also very, very excited. Then, at some point, I will just go up to her, roughly yank her robe apart (or her panties down, or pull her skirt up -- this has happened quite a bit), bend her over the stove and take her right there. David, she LOVES it.

Even when she says she doesn't want to have sex (she doesn't like to immediately after eating a big meal), if I feel like it, I will tell her, "I don't care," and I will slowly, slowly, slowly unbuckle my pants and take them off and walk towards her. Initially she runs, and I just walk slowly behind her and follow her. Of course, before too long she can't wait for me to catch her and bend her over whatever piece of furniture happens to be close by. At this point, she usually has her first vaginal orgasm within 30 seconds, her second within another minute or two and she will always have a third when she can feel that I'm getting ready to cum.

The point is, David, I am having the most incredible sex, as often as I want, and of course, I am giving her the best sex of her life and she can't get enough. Do you want to know the one thing she always asks me? She'll look at me after I've just taken her, or given her an orgasm over the phone, or told her that I was going to fuck her in her office, or anytime I am direct and tell her what I'm going to do? She looks at me expectantly and asks, "You're never going to change, right? This isn't some sort of trick? You'll always be this way, right?"

She had two orgasms the other night before I even got in her all the way. I just stuck the head in and let it sit there and she had two - just from the head.

This is, of course, a hot girl that is used to having men pursue her and wine and dine her, take her on private airplanes to movie premiers, etc, etc, etc. She used to work as a bartender in NYC, so she knows all the moves. Now she spends all her money to come and see me, blows off her friends and family and just wants to come here on the weekends and let me fuck her all day long.

Her favorite place for me to take her -- the grocery store, because then she knows that we're going to stay in and she'll get more time in bed. Dude, I'm fucking serious. Her favorite place to go -- our "date" place is the grocery store.

On the weekends, I generally fuck her at least 3 times a day and she usually has at least 3 or 4 orgasms every time. Seriously, I don't think there has been a day in the last two months where she has had less than 15 orgasms in one day. Sometimes, when I'm feeling rowdy, I'll put her in a position that I know she can't help but have orgasms in (hitting the deep spot) and make her continue to have orgasms until she begs me to stop.

The only bad part about all of this is that she really doesn't have anyone she can share this with. None of her friends would believe that she has continuous orgasms that last 15 minutes or more, none of them would believe that she had over 20 orgasms in one day -- so she feels a little frustrated that she can't actually share this with anyone.

Still, truth be told, she was ready for this when I found her. She just needed someone to lead her -- and truthfully, it's been very easy. She fits the profile you indicated about being a Daddy's girl and highly intelligent (she's a Mensa chick), so you were right on the money about making sure you choose the right one.

David - this is all from post-coital suggestion and making her believe that everything I say is going to happen WILL happen, and then it does.

Next I'm going to work on making her cum as soon as I put my dick in her and then the "orgasm on command".

David, life is good. Life is oh, so fucking incredibly good.

Thanks for showing us the way.

Robert, you chose wisely, and you were man enough to bring out the woman in her.

Some readers may be taken back by some of the things you said to her, such as "I am going to fuck you as often as I want and wherever I want", but it was said in the specific context of respect, enhancement of the relationship you two share together, a celebration of her sexuality, and understanding the fact that she loves to be a woman. This is evident in her reply, which was basically, "Promise?"

Here is an e-mail from George in Bulgaria:

I am 25 and I had so little experience that I considered myself a virgin. So I started learning about women and pick-up. That gave me some confidence but I felt it wasn't enough. Until I came to you.

When I met my girl she told me about her experience, and to be honest I felt a bit intimidated. She had far richer experience than I. She told me that she had had only two orgasms - one clitoral and she wasn't sure what the other was. Though I felt self-doubt, I knew that I knew stuff none of my male friends knew about women. That gave me confidence stronger than my fear.

What I did first was to play with her deep spot and G-spot, and YES - I gave her her first vaginal orgasms! Man, was she grateful! She turned into this wonderful sexual creature that had "screaming orgasms" every time we had sex. She even calls me "Lord of the Fingers" from time to time. At first I thought, "What the fuck?!" I felt like THE MAN.

Now, about last week. She came to stay with me for the holiday week. You say it is important to use dirty talk, so I decided to try it. I didn't know what the response would be. So during sex I just told her, "You love when I fuck you like that, don't you?" It was like she was WAITING to hear that. "I love when we have sex this way" was replaced by "I love when you fuck me really hard from behind". (She loves the doggy style position, which she first tried with me.) Now she says "have sex" less often than "you fucking me". Once she said that she wants me to tie her hands and "rape" her, and once "I've never been fucked that good before". It's like I opened the floodgates of her sexuality.

For the last 4 or 5 days I felt like I was living in a porn movie. She was doing things she'd never done before, she was jumping on me and ASKING me to "fuck her hard" while I was reading a book or watching TV or she sat on my leg and started humping it.

I know that there are men who ask their partners whether they came during sex. If I ask my girl it won't be "whether" but "how many this time".

Once I was stimulating her clit, using your

Welcomed Method. It became so sensitive that I was only gently rubbing the shaft and she came powerfully. She said she felt "little needles" all over her body, even on her face. From then on every orgasm was AT LEAST that powerful, and she squirts every time! Once she said she came four times during the foreplay alone.

She told me that sometimes I turn her on just by looking at her. One time we were watching TV, my leg was next to hers and I was moving it up and down unconsciously thus rubbing hers lightly. She turned and just said, "Stop it, you're turning me on."

The techniques I use - ONLY G-spot and deep spot stimulation and sometimes clitoral. Really, just that. More important though is the mindset I've developed. I TELL her what to do and she loves it, I LEAD her and she loves it, I call her "my little slut" and she loves it.

When she has an orgasm and squirts I comfort her by telling her, "That's my girl," or, "Good girl." Most of all I LOVE caressing, hugging and kissing her after her orgasms. I LOVE to do that. It's a wonderful feeling seeing her so satisfied and so happy while she is in a trance-like state or smiling or giggling or crying trying to wind her body around mine like a snake or trying to touch every part of my body.

There were times when I came 3 minutes after sticking it in but now I feel like I can control when I come and I don't care if I come or not, really!

I realize that for many guys this will seem like I made it up and I didn't quite believe it at first... but it is true. I even don't have time now to describe everything thoroughly!

P.S. Are all women like that or did my girl

just turn out to be hypersexual? David, after you, picking up girls seems like child's play.

Thank you for the person you are and for what you do!

You truly change men's lives, mine for sure!

Such fantastic success from simple techniques, dirty talk, and the right mind set.

Here is an e-mail from Russell in Indiana:

I'll start with my experience before reading your books. I knew only what I had seen in the movies, as far as how to have sex. Then I met a very experienced girl (a stripper) and she taught me a lot about how to make a woman feel good, and she enjoyed everything. I had a lot of fun with her.

Then I met my future ex wife. She was also very experienced, but she had never had an orgasm. I thought I knew a lot about women, and I told her I thought I could give her one, so I tried doing oral but it just tickled her. I pretty much tried everything. Then we bought her a vibrator and she had her first orgasm. It was neat to see the look on her face. She was actually able to have many orgasms with this vibrator, so that's how we had sex every time.

I actually gave up on trying to give her one without it. I just thought that was the only way she could have one. So 5 years and 2 kids later, she left me and we divorced. It crushed me beyond belief. I know you know what it's like, but the difference was she got the kids and wouldn't let me see them.

After 6 months of feeling sorry for myself and hardly talking to her, I decided to e-mail her and we started talking. In one of her e-mails, she was being a bitch and told me that she had had orgasms with other guys without a vibrator. That really hit me in the gut. I asked her why she didn't tell me what she liked, or what felt good, so that I could do that. I, like every other guy, don't what to be considered lame in bed.

So I began my research and started reading everything I could on sex and giving a woman an orgasm. I wasn't really learning anything until I came across David Shade's "Give Women Wild Screaming Orgasms". I want to say God bless you for writing these books. I studied it over and over. Now I felt like I really knew something.

I bought your other two e-books also, and I'm still studying them. It really has boosted my confidence, and I'm a lot less shy around new women, where before I would hardly say hello to them.

I also found a woman who is worthy of being with me. She was the second one I got to use my newfound knowledge on. The first time we had sex I could tell I had blown her mind, and it made her pussy sore from having so many orgasms (4). She has been up for pretty much anything since that first night of having sex.

She didn't like giving BJs and she didn't want anything to do with anal, but after reading your "Manual" and giving her many, many orgasms, she let me do anal on her. At first she said it hurt really badly, but I told her, "I know baby, but you like it; you like the feeling of me being inside your ass." A few moments later, after more dirty talk, she had one of her most powerful orgasms. She was dripping from her pussy. We did it 3 more times that night and every night for about a week.

I also started planting other seeds in her head. I used your "Now She Loves To Go Down" chapter from your Manual. She never even liked giving head at all, and before I knew it, she was giving me head all the time. I would always plant seeds in her mind about how she liked sucking my big cock, and how it turned her on. Then one night she kept on going with it (before she would do it only for a few minutes). I told her I was going to cum. She swallowed it all, even squeezed out everything, licked her lips, and said it tasted good.

I couldn't believe it. Giving women pleasure is the most fun I could have imagined, and she tries to make sure that I get what I want and that I feel good. I love it.

She is up for anything. I haven't gotten as far as the hypnosis yet, but I look forward to it, and she does too.

I want to thank you again, David. You have really helped me beyond belief. Keep everything up, and I look forward to any more books you write. I'll be your customer for life, and I recommend your books to all of my friends.

There is a tremendous amount of pent-up sexual potential in a woman, and men need to understand the immense amount of power that men have because only a man can release that potential in a woman.

That requires BEing a Man. A man who can drive like a man. A man who can lead. A man who adores everything about her that defines her as woman. A man who revels in bringing out her ruthlessly expressive, decedent, primal, carnal, insatiably hungry natural woman in her.

Here is an e-mail from Bob in Palo Alto California. I consulted

THE SECRETS OF FEMALE SEXUALITY

him on issues of dating and sexuality.

Unleashing the raging sexuality in a shy, incredibly sexy, virgin daddy's girl:

When I met David Shade about 3 years ago he immediately struck me as being a personally powerful man. When he spoke, it was with an authoritative tone that said, "I know what I'm talking about, and if you truly want to go down the path, I can show you the way." On the other hand, I was a selfish kid with little experience with women and had no idea what much of what he said meant at the time. I say selfish because deep down I wanted to learn how to seduce and sleep with attractive women for my own self-validation, which I didn't have from any other sources. He will show you the path, but you must walk down it.

Looking back, I remember sitting over coffee, or lunch, and he'd put questions to me like, "What is it that you really want?" "What if you had exactly what it is you wanted? How would that be for you?" "Why would a beautiful woman want to sleep with you?" "Why do you want to fuck hot women? What would be true for you?" I kept thinking, "Why wouldn't my mentor give me a straight answer? Why was everything he said so cryptic to me at the time?" Still, I thought about what he said a lot. He was testing me. Testing my character and my ability to think for myself.

Through spending time with David, I soon began picking up pieces of his belief system, a little bit at a time. I learned from listening to him and watching him work in the real world about being a personally powerful man.

I learned firsthand what he is talking about when he says "self-respect above all else". I learned about how women are amazingly sexual creatures, and how the amount of potential in a woman's raw sexuality is beyond what most men will ever know. The right women of course. High self-esteem women. I learned about how low self-esteem women will actually not allow themselves to be completely sexually aware, because they feel they don't deserve it. I also learned that the high self-esteem women are the ones with the most potential, because deep down locked inside them is a box full of amazing sexuality and dreams and passions. David gives you the keys to unlock that box.

I stopped worrying so much about sleeping with women for my own selfish gratification, and started wondering about how hard I could make that hot girl in line at the store come if I took her home. I started wondering if she was capable of letting me show her more about her own body than she knows is possible... I started wondering about giving her the gift of unlocking her box of sexual fantasy and pleasure. In short, I really started appreciating the sexuality of women a lot more. I learned that you can never make a woman have sex with you. You can never convince her, and if you do, she won't be enjoying it and is doing it out of guilt. What if she was in the state of mind to WANT to give you the most amazing pleasure of your life... what if she was in the state of mind to give you full access to her body because you can make her feel better than anybody else? What if she genuinely wanted to give to you everything possible, because you give her ground-shaking orgasms every time you are together?

So, I went and found my current girlfriend. I'll save you all the pickup stuff. There are plenty of guys out there that will teach you the arts of pickup. This will focus on what to do after you HAVE the girl. When we first met, she was a shy daddy's girl... incredibly sexy. This is the kind of girl that when you are walking down the street with her on your arm, other guys look, and they look often.

She was a virgin when we met, and really had no sexual experience. When I first kissed her she couldn't even kiss me back with any sort of intensity because she was so shy and nervous... She would just blush and giggle and slowly kiss me back, yet she liked kissing me... With my newfound belief systems I learned from David Shade I thought... how wonderful! She's never even had the misfortune of being with a self-validating, one-pump chump, and I want to make her come so hard that she shakes before I even give myself to her! She was very, very shy, but through our conversations I could see she was very high self-esteem. She was very intelligent, she was very close to her parents, she cared about ambitions and goals in life. The more we talked the more I started to see her sexual potential in her words... The closer we got, the more our talk evolved into some of her sexual fantasies.

Here was a beautiful, amazing, sexual girl, yet it was all trapped inside her, and she was sitting waiting for a man to come along and show her how to let it all out.

I started with some basic things on the phone... First I would let her describe an ideal sexual situation -- ones she'd fantasized about since she was a teenager... Then I would tell her, "Mmmm baby I love how wet your little pussy gets for me," as I heard her moaning into the phone with pleasure. I started getting her to imagine these situations so vividly that she'd have orgasms on the phone with me. There were a few times when we'd get together and I'd only take it so far with her -- up to the points I had described to her and let her already imagine... Then I'd let her imagine going further, and after she could vividly imagine that I would physically take her further, I'd say, "Can you see that in the picture baby? Can you look down and see that hard cock sliding in your pussy?" Once somebody visualizes something in their mind it is much more real to them.

Soon she was having powerful orgasms for me and telling me how good she came. She also became much more verbal with ME... She started telling me how bad she wanted to do all these things to me, and would describe them to me in great detail for me.

David taught me what women really want: They want to be treated with dignity and respect -- when they are around other people in the world and out and about... but she wants more than anything in this world to let that raging sexuality out of her and have her man fuck her like a naughty little slut when she is in the bedroom with him. Good sex to a woman is the physical manifestation of an incredible emotional bond with the man she loves.

I started telling her how I love that she is such a nice proper good girl around everybody else... mmmm, but when you are with me, you can be a naughty little girl, can't you? She would say, "Oh yes... I love being your naughty girl, but only yours!!!"

When we did actually get intimate the first time, all the talk and preparing her mind paid off -- big time. First, I started with foreplay and touching her all over until I could feel her pussy dripping wet for me. Then, I used David's techniques mentioned in his "Give Women Wild Screaming Orgasms" book on giving oral sex to your woman. If you don't know the anatomy of a woman's sexual organs you are really doing yourself a disservice in not having David's book! First I licked her clit steadily with my tongue until she had her first orgasm, then I licked her clit and gave her the deep spot at the same time, which drove her completely insane!!! I also used variations of other things in the "Give Women Wild Screaming Orgasms" book, like the hummer. I was in wonder lying there licking her clit, pushing my finger into her and giving her the deep spot and seeing how amazingly sexual my woman could be, and how I was unlocking her box right in front of me. She moaned, "Oh god oh god oh yes oh yes, I'm cuming for you!!! "

I realized at that moment there is nothing quite as rewarding as hearing your woman moan with intense pleasure and looking up at her face after she came so hard for you that her legs are quivering and seeing a big smile on her face. After that, she practically begged me to fuck her, which I did until she came again, soaking our bed sheets from her wetness by this point. I whisper in her ear how good her little pussy makes me feel when I fuck her, and tell her how she is MY girl, and I love how nasty she can be with only me. The more I intensify those feelings the wetter she gets, and the harder she gets off for me. It really is quite staggering.

These days our relationship couldn't be better. I am exploring new things with her sexually all the time, and she loves every bit of it. Guys, once you give your woman incredible pleasure like she has only dreamed of, she will want nothing more in this world than to make her man so happy and for him to come for her over and over again. Do what David tells you, be the man, learn about how to give your woman incredible pleasure because you love women, you adore women, and you are fascinated by their sexual capability... Then give it to them, and I assure you, your end will be taken care of. One of a woman's biggest turn-ons is how she can make the man she loves feel sooooo good, and she will want nothing more than to do that for you after you enlighten her in ways that David teaches you how to do.

Today my girlfriend told me how much she is looking forward to exploring all of her naughty little sexual fantasies with me, and wondered if I'd mind her learning them with me. :) I reward her good behavior now with things that she TRULY WANTS to do for me. When she tells me a sexual fantasy of hers I'll reward her by telling her that, yes, she's been a good girl for me and I will LET her enjoy those things. For instance, today she was telling me how she wants to try deep throating me while I lie back and enjoy it until I explode, and I said, "Mmmm yes baby I promise you that next time we are together, I will let you do that for me." When you are a sexually powerful man, and you show your woman pleasures that she's only had in her dreams, you will have full access to do anything you want to her, and you will have that access because she truly will long to experience those things with you.

When you learn David's material, your woman will beg you to do things to her that other men have to try and convince their women to do! I come all over her body because she asks me to now, I put my cock in her mouth, have her stick out her tongue and finish in her mouth... because she wants to make me feel so good -- then she swallows every drop with a big smile on her face. I spank her and tell her how I love her to be my naughty little girl tonight, and she goes crazy and starts moaning. She already has a little list of positions she wants to try next time we're together. In short, when you take care of your woman the way she needs to be taken care of, she will literally obsess about you because that is what she has wanted whether she knows it or not since she was a little girl.

David Shade has been a profound influence on my way of thinking, and as a result a huge influence on my life. Let him show you the path, and then let yourself walk down it, and see for yourself the enjoyment of truly giving women incredible pleasure, and reaping the benefits of doing that for her!

Bob's success began at the beginning, when he chose wisely. Bob and his girlfriend are now engaged to be married.

Find a woman who wants to enjoy everything that is woman, and then BE the Man.

Here is an e-mail from Randy:

I am writing to you so that I may share with you the success I have had with your Manual, and most importantly, thank you for being such a positive influence on my life.

I have wanted to write to you and express my thanks for all your help, but I was putting it off until I could write to you about succeeding in doing something that really caught my attention in your Manual.

I met a woman that passed the tests you taught me about being selective to find the "good ones".

We went out for some coffee at a place that I knew closed early (exit strategy if I did not like her, and reason to go back to my place if I did). We were back at my place in a bit under two hours, were kissing an hour after that, and she was having her first orgasm with me about an hour after that.

The interesting thing about this one is that her first two orgasms happened while she was fully clothed. The first orgasm left her with the most amazingly beautiful look of utter shock on her face! She kept saying, "How did you do that? Oh my god, how did you do that?"

At this point, she wanted to get fucked big time. She was pulling at my clothes, but I made sure to remove hers first, and then rub her clit briefly using a simplified version of the Welcomed Method. This drove her wild, and she said, "Oh my God, I can't even find it that fast!"

I wanted to be sure she would be vaginally orgasmic before having intercourse, so I got her close and then slid my finger inside her and rubbed the front of the Deep Spot. Boom! She came really hard!

I knew I had the green light at this point, so I gave her a moment to come to grips with all these new experiences, took the rest of my clothes off, worked her up with some sexy dirty talk, slapped on a condom, and with only the slightest bit of thrusting she came like crazy!

I was having some serious fun, so I went ahead and gave her a couple more, even though I had not ejaculated. She was not going to let me get away with that though, and so she rolled me over and went down on me. I know she was not validating herself by doing so. She truly wanted to reward me for helping her feel sexier than she had ever felt before, and since she was so wonderfully responsive in every way to me all night long, I was happy to show her how good I could come for her.

I decided to progressively train her, seeing her approximately one night a week so that she would have sufficient time to anticipate seeing me and drive her to fantasize about me.

The second night I started things off with two more Think-offs and then a nipple orgasm.

I then teasingly chastised her for

fantasizing about me and showed her how she took "this hand" and "put it right here" and taught her to masturbate in front of me. We then had intercourse and I lost track of the number of orgasms.

The third night I did the Welcomed Method on her and taught her to relax into her orgasm and sustain it.

The fourth night I did the humming clit sucking and that of course worked wonderfully. She was fully convinced that anything I said would happen would, indeed, happen.

I did her doggy while holding her hair and talking dirty. I've always fantasized about doing a woman doggy, and any time that I did in real life, I would come almost instantly because it was such a powerful fantasy for me.

With the dirty talk, it puts my focus on the woman and her experience, which allows me to screw like crazy without getting overly focused on my own pleasure and ejaculating. Having a woman in orgasm while I did her doggy was like climbing Mt. Everest and finally making it to the summit; it was a huge success for me.

The next night I did a ten count to orgasm, and I was actually a little bit unsure how she would respond, but she didn't skip a beat and came right on ten!

It was a weekend, so she stayed the night and the next morning I woke her up with her on her stomach and some morning wood pressed against her ass. I get her turned on right away and she must have still been wet from the night before because I slid right in with very little foreplay. I told her to reach under and rub her clit while I fucked her. You know the results.

I constantly prove to her with my actions

that I am fully vested in her pleasure and that I am also interested in ALL of her emotions. I do not rush getting to sex, I watch her to signal me, and then I will even slow her pace down to build anticipation. I usually make her talk to me after sex.

Also, I have never formally induced trance in her since she proved to be so highly suggestible from the very beginning. I decided that the experiences would be more powerful for her if she could not try to convince herself that she was having the experiences only because of hypnosis.

She eventually told me that before meeting me she had a difficult time even having one single orgasm during sex. I've been nurturing her fantasies and she told me that she used to masturbate to the imaginings of movie stars or men with no faces; never someone she knew. She now regularly fantasizes about me when she masturbates.

My lovely and willing love lab test subject was over at my house, but she was on her period. I could just tell that as much as she would prefer otherwise, she was expecting me to be inconsiderate of her situation and uninterested in her until her 'curse' had passed. I saw this as a big opportunity!

I knew she could orgasm without me even taking her clothes off, so I proved to her just how sexy she could remain (to me and to herself) even during that 'time of the month'. I did my little "magic trick", made her feel wonderful, and then explained to her that orgasms while menstruating massage her reproductive organs, helping to eliminate pain and cramping.

She hugged me tight and was taken with gratitude for me doing something entirely for her benefit. Of course, I did not tell her that it was part of a larger process that she was unknowingly going through. I wanted her to have an orgasm while sucking my cock.

Fast forward to one month later (last night) and she is on my couch, anticipating some affection. I've been developing her trust and responsiveness for some time now. I had held back Think-offs for a while, because I knew that I wanted to keep them fresh and exciting and powerful for this moment.

I gave her an orgasm fully clothed and she was just as taken by it as ever. Then I took off her shirt and she went right ahead and took off her bra like a good girl. I proceeded with a nipple orgasm that had her once again in utter amazement.

I then got the dirty talk going dirtier and naughtier, getting her excited about going down on me. She THOUGHT it was my turn to get off, but I had other plans (hehe). She went at it and I went to town with the dirty talk and progressed into more and more suggestions about how horny my cock made her and how sucking me made her clit so hard, etc. She of course continued with her 100% success rate of feeling everything I tell her to feel and had a powerful orgasm while sucking my cock!

She knocked my wood down enthusiastically right after and then commented on how that was the wildest thing she has ever done! I want to revisit that experience, so I made the suggestion that she was going to have a lot of fun fantasizing about what just happened. She agreed, thus setting up regular rehearsal in her mind, making it a total reality for her.

Thanks man, I'm finally living my life, and you helped me make it happen.

The reason why she came so readily and often for you is because she is highly emotionally moved by you. You command respect with her. She takes what you say and do seriously. I'm not surprised she keeps coming back for more.

It is much more than "rub here, rub there". It is very much emotional and mental. You truly understand that Masterfully.

The reason why you don't even need to hypnotize her, and why she is so highly responsive to you, is that she completely trusts you. Trust and respect, in the context of a relationship, are very much related.

Being a Masterful Lover has nothing to do with experience. It is all about knowledge and beliefs. Even a virgin can do it. Here is an e-mail from Nathan in Vancouver:

I'd like to share a success story with you, one that shows how your writings have helped every part of my love life.

First off, before I'd come across your newsletter, I was a virgin (though I'd dated and been in relationships a bit) who had become obsessed with trying to learn pick-up techniques, etc., but coming from a place of insecurity. I thought that I needed to change myself into a bar star horn-dog who could and would pick up a dozen chicks in a night. Not surprisingly my success rate was low and my motivation was equally low. I knew that there was something wrong with my game.

One day, I came across you through an interview of you and was at first totally put off by this man who talked about women being dirty, sexual beings. However, after a few weeks I listened to the interview again and was blown away. I immediately signed up to your newsletter and, by chance, found a friend who was willing to lend me his copy of your Manual. In very little time, the concept of a Masterful Lover just clicked in my head, like it was the missing piece of the puzzle.

I had a girlfriend at the time I discovered your work but I had messed things up due to all the crazy ideas about seduction I'd had earlier so I cut her loose without any fear of loss. She appreciated my honesty and we're still friends.

Within 2 days of the breakup, I found and began to date an amazing woman (who I found thanks to my newfound confidence). I've been with her ever since and have kept blowing her away with dirty talk, fantasies, orgasms of all types, ultimately respecting every aspect of her femininity. As a result, she is always telling me how she feels so safe with me, how she wants to do crazy sexual things with me that she never used to be interested in, and how I make her feel like a woman. All this from a girl that was a virgin before she met me!

I truly feel like THE man now that everything about my love life has improved (the meetings, the emotional closeness, the sex, and even the breakups!), and it's due to you.

So thank you, and feel free to use this as a success story in your e-mails or as a testimonial for the power of your writings.

And here is another success story from Osagie:

I was a 28 years old virgin until 2 months ago. My celibacy was born out of religious reasons and some limiting beliefs, but that's a story for another time.

After much introspection at the beginning of this year, I decided to pursue a relationship with a girl I was friendly with. Although we often kissed and made out, we never had intercourse. There were actually subtle cues from her indicating she wanted to take things further sexually, but I was more than aware of my sexual ineptness and performance anxiety. I knew if I wanted to continue to keep her, I would need to arm myself with the necessary skills regarding making love to a woman.

Fortunately, I came across your material while rummaging through information on the Internet and immediately signed up for your newsletters. The newsletters were thought-provoking, prompting me to purchase your "Give Women Wild Screaming Orgams" e-book.

Without mincing words, the information contained in your foundations e-book was like dynamite shattering all my limiting beliefs on sexual intercourse. One thing that really impressed me about the book is that you cut through all the fluff, presenting only the meat on the subject.

Knowledge, they say, is power! Armed with the facts from your book and a newly found confidence I finally had intercourse with my girlfriend. The experience was ethereal. Even Mr. Webster can't describe how beautiful the experience was. Like you taught in your book, I led in the bedroom and she followed. She practically browned out when I applied the Welcomed Method on her and applied the deep spot technique. In fact, at a point she was so overwhelmed, she started panting like she was in a trance. I had never seen anything like it. I must admit I was a little scared until she held onto me and said she had never experienced anything so pleasurable until now!

In retrospect, I have come to realize that being good in bed is not about experience or how many lays you have under your belt. Rather it's all about knowledge and skill.

I can't thank you enough.

SECTION 4

Resources

There are only four questions of value in life Don Ottavio. What is sacred? Of what is the spirit made? What is worth living for? And what is worth dying for? The answer to each is the same. Only love.

- Don Juan DeMarco

Every true lover knows that the moment of greatest satisfaction comes when ecstasy is long over, and he beholds before him the flower which has blossomed beneath his touch.
- Don Juan DeMarco

Resources

The next step that you want to take is to study my program "Give Women Wild Screaming Orgasms". This includes the correct knowledge and beliefs to being a Masterful Lover, sexual techniques for giving women incredible pleasure, being comfortable with her sexuality, and developing the enabling beliefs for sexual confidence.

WildScreamingOrgasms.com

In my audio CDs, I tell you how to use sexy dirty talk to drive her wild with excitement; "The Art of Sexy Dirty Talk". It also comes with a bonus CD in which I tell you "one step at a time" how you can use a woman's fantasies to lead her to become a ruthlessly expressive sexual being; "How To Set The Foundation For A Wild Sexual Relationship".

SexyDirtyTalk.com

If you want to actually hear David Shade talking dirty, you can hear it in "Give Women Hot Phone Sex". Three audio CDs of actual recorded phone sex episodes, where you will hear me talking very dirty, and you hear the actual female screaming orgasms that result. You will discover, by example, exactly how you can make your woman come with just the sound of your voice, and keep her coming for as long as you like!

GiveWomenHotPhoneSex.com

Finally, "David Shade's Manual", hypnosis for couples. This is the really advanced far out stuff for using NLP, hypnosis, and energy for giving women indescribable pleasure. Some would even call it kinky.

DavidShadesManual.com

In review: Start out with "Give Women Wild Screaming Orgasms." Get good at the sexual techniques and develop your sexual confidence. Then move to the audio CDs and become good at using sexy dirty talk. Then move to "Give Women Hot Phone Sex" to totally open herself to you sexually. Then get into the really advanced stuff with the "Manual".

For further reading, check out my articles on my blog at:

DavidShadesBlog.com

All this and more can be found at my site for developing yourself into a personally and sensually powerful man.

RenegadeSexExpert.com

I also include the list of books and mentors that have affected me along the way. I have consumed countless books and resources on this topic, and can recommend the few that contributed leading work

"Dangerous Men and Adventurous Women" Edited by Jayne Ann Krentz, c 1992 by University of Pennsylvania Press, 186 pages. ISBN 0-8122-1411-0. Mark Cunningham suggests we read this. It is a collection of essays by 19 leading romance authors on the appeal of the romance novel. Describes all the traits of the romantic hero.

"A Passion For More" by Susan Shapiro Barash, c 2001 Berkeley Hills Books. ISBN 1-893163-24-5. The stories of 57 women and what drove them to have extramarital affairs. In almost every case, the woman's thoughts are consumed by the man with whom she has the exciting sex.

"Private Thoughts" by Wendy Maltz & Suzie Boss, c 1997, 2001, ISBN 1-57731-146-9, first printed in 2001. This book presents an outstanding thesis on the why and how of female fantasies.

"Story of O" by Pauline Reage, c 1954. ISBN: 0345301110. It is a true story written by a woman under a pseudonym, but whose identity was revealed in 1994 as Dominique Aury. Considered the classic reference for BDSM.

"9 1/2 Weeks A Memoir of a Love Affair" by Elizabeth McNeill, 1978. ASIN: 0425103846. It is no longer in print, and used copies in good condition sell for ten times the original price. It is a true story also written by a woman under a pseudonym, but whose identity has never been revealed.

"Different Loving: The World of Sexual Dominance and Submission" by Gloria G. Brame, Jon Jacobs, Will Brame, c 1996, ISBN: 0679769560, Publisher: Villard Books.

"The Erotic Mind: Unlocking the Inner Sources of Sexual Passion and Fulfillment" by Jack Morin, c 1996, ISBN: 0060984287, Publisher: Perennial.

"Arousal: The Secret Logic of Sexual Fantasies" by Dr. Michael J. Bader, ISBN: 0312269331, c 2002, Publisher: Thomas Dunne Books.

"My Secret Garden" by Nancy Friday, c 1973, ISBN: 0671742523. The documented fantasies of hundreds of women, in intricate detail. Simply fascinating.

"The Hite Report" by Shere Hite, c 1976, ISBN 0-020551851-8. A huge study of female sexuality. Read how sexually frustrated women truly are.

"Think and Grow Rich" by Napoleon Hill, c 1960, ISBN 0449214923. The greatest book ever written. Hill worked with and studied the great captains of industry, such as Carnegie, Edison, and Ford, and documented all the steps to success. His book is to billionaires what Machiavelli's "The Prince" was to kings. Apply the teachings of this book towards female sexuality, and truly anything is possible.

Get Your FREE CD Today!

Receive a FREE CD in the mail!

In this one hour interview, David Shade

reveals to you the secrets of...

How To Keep The Relationship

Exciting!

For details and to receive your Free CD

go to my website:

RenegadeSexExpert.com

FREE VIP Inner Circle Offer

Special FREE Gift from the Author

Test Drive David Shade's

VIP Inner Circle Membership

for 39 days FREE

- VIP Inner Circle Newsletter
- Exclusive Audio Interviews

Get Your FREE Gift Here:

Masterful-Lover-VIP.com

There is a small one-time charge to cover postage for the free VIP Inner Circle Membership, and you have no obligation to continue at the low monthly rate. In fact, should you continue with membership, you can later cancel at any time.

FREE Stuff Linked To This Book

To help you get full value from this book,

there is a collection of

FREE EXTRA RESOURCES

waiting for you at

RenegadeSexExpert.com

FREE e-mail course "Secrets to Mind Blowing Sex!"

FREE teleseminar with David Shade

FREE access to ask David Shade your

most burning question

Printed in the United Kingdom
by Lightning Source UK Ltd.
128397UK00001B/383/A